Theatre in Co-Communities

Theatre in Co-Communities

Articulating Power

Shulamith Lev-Aladgem

First published 2010 by
PALGRAVE MACMILLAN

Palgrave Macmillan in the UK is an imprint of Macmillan Publishers Limited, registered in England, company number 785998, of Houndmills, Basingstoke, Hampshire RG21 6XS.

Palgrave Macmillan in the US is a division of St Martin's Press LLC, 175 Fifth Avenue, New York, NY 10010.

Palgrave Macmillan is the global academic imprint of the above companies and has companies and representatives throughout the world.

Palgrave® and Macmillan® are registered trademarks in the United States, the United Kingdom, Europe and other countries.

ISBN 978–0–230–55519–8   hardback

This book is printed on paper suitable for recycling and made from fully managed and sustained forest sources. Logging, pulping and manufacturing processes are expected to conform to the environmental regulations of the country of origin.

A catalogue record for this book is available from the British Library.

A catalog record for this book is available from the Library of Congress.

10 9 8 7 6 5 4 3 2 1
19 18 17 16 15 14 13 12 11 10

Printed and bound in Great Britain by
CPI Antony Rowe, Chippenham and Eastbourne

*To my dear husband Dr David Aladgem (MD)
and dear daughters Gal and Shiri*

Contents

Acknowledgements

Parts of this book have appeared in academic journals. I thank the journal publishers for permission to use this material.

'Improvisation upon the Scroll of Esther: Symbolic Inversion in an Adult Day-Care Centre', *Journal of Folklore Research*, 35(2), 1998: 127–45.

'From Ritual to Drama and Back in a Day-Care Centre', *Journal of Aging Studies*, 13(3), 1999: 315–33.

'Dramatic Play amongst the Aged', *British Dramatherapy Journal*, 21(3), 2000: 3–10.

'Carnivalesque Enactment at the Children's Medical Centre of Rabin Hospital', *Research in Drama Education*, 5(2), 2000: 163–74.

'From Object to Subject: Israeli Theatres of the Battered Women', *New Theatre Quarterly*, XIX (Part 2), 2003: 139–49.

'Whose Play is it? The Issue of Authorship/Ownership in Israeli Community Theatre', *Drama Review*, 48(3) (T. 183), 2004: 117–34.

'The Israeli National Community Theatre Festival: the Real and the Imagined', *Theatre Research International*, 30(3), 2005: 284–95.

'Between Home and Homeland: Facilitating Theatre with Ethiopian Youth', *Research in Drama Education*, 13(3), 2008: 275–93.

All translations from Hebrew are by the author.

Introduction

This book has grown out of a long journey that began when I chose to leave professional acting for an academic career, and which has focused to a great extent on facilitating and researching theatre within marginalized communities in Israel. The early projects mainly involved facilitating theatre among the disabled elderly at geriatric day-care centres. My major goal was to provide them with the appropriate theatrical modes by which to articulate their feelings and thoughts. As I explored the therapeutic potential of these projects to make a difference in the daily lives of the participants, my focus broadened to include other marginalized groups. Over the years I have been engaged in researching community-based performances produced by Mizrahi Jews (originating from Muslim or Arab countries), Jewish-Ethiopian youth and Israeli-Palestinians. This research raised additional issues, such as nationality, ethnicity, class, gender, identity and social change.

The course of my ongoing work relates to the striking direction theatre has taken since the second half of the twentieth century. With its expansion from the centre of society to the periphery, theatre has been appropriated by various marginalized communities and reinvented as a means to a cultural intervention that combines art with social action and aesthetics with pragmatism. This wide cultural phenomenon has acquired a broad range of names, from applied theatre, community-based theatre, theatre for development, popular theatre and social theatre to theatre of the oppressed.[1]

In this book I use the expression 'theatre in co-communities' as a comprehensive term that can serve as an umbrella for all the others.

1

'Co-community' is my adaptation of the more general term 'co-culture' suggested by the communication scholar, Mark Orbe, whose intention was to avoid the negatively charged prefixes of words such as *sub*ordinate, *sub*cultural and *non*-dominant that traditionally describe minority groups (Orbe, 1998, p. 1). Whereas 'co-cultures' indicates the complex network of minority groups/cultures that constitute Western society, the more specific term, 'co-community', refers to a particular, local part of a larger co-culture. Theatre in co-communities thus refers to the theatrical articulations of groups of local, non-professional performers who, through the theatre projects, represent not only themselves and their local co-communities but also the greater co-cultures to which they belong.

Although Israel, a democratic Westernized state, was the setting for the work that is the subject of this book, some of the co-communities explored here, such as the disabled elderly members of a rehabilitation day-care institution and hospitalized children in a medical centre, are also typical of other societies. Other co-communities discussed here, such as the Mizrahi residents of an underprivileged neighbourhood, young Ethiopian students at a boarding school and the Israeli-Palestinian inhabitants of a mixed-ethnicity town, are more specific to Israel and reflect its unique social construction. Israel, as Sami Smooha (1978, 1989, 1993), a leading critical sociologist indicates, is an 'ethno-democratic' state, heterogeneously comprising various ethno-cultural groups within a single state. The local particular mode by which these groups are integrated is differential, in that each group has a different status in the overall societal frame. The dominant group is that of the secular, middle-class and veteran Ashkenazim (Jews originating from Europe and America). Other Jewish groups that immigrated to Israel after the establishment of the state in 1948, such as the Mizrahim and the Ethiopians, are the subaltern groups that have been assimilated into the dominant culture through paternalism and co-optation. While these co-cultures are formally included in the Jewish national state, they are effectively considered as 'the stranger within us' (Shenhav, 2003), and the Palestinian citizens of Israel who are generally perceived as 'the total strangers' or 'the total other' are basically not integrated but nonetheless dominated by the state. In addition, the political culture in Israel has become more complicated over the years as a result of the contradiction between the two different discourses that the state seeks to pursue. On the one hand

Israel endeavours to fulfil the inclusive principle of democracy while on the other hand it attempts to follow the exclusive principle of Jewish nationality. Caught between apparatuses of equality and discrimination, segregation and integration, universalism and nationalism, Israeli society has suffered from severe cleavages, mostly those between Ashkenazim and Mizrahim and Jews and Palestinians (Peled and Shafir, 2005). Although this book discusses the complexities of Israeli society from the standpoint of its co-communities, the diverse collection of case studies also engages with cross-cultural issues such as power relations, identity formation, gender, class, ethnicity, nationality, representation and empowerment. These topics are of interest for theatre facilitators and scholars throughout the world, who will find in these chapters an echo and a dialogue with their own experiences.

The book presents my own work too, as a theatre facilitator and researcher, with the intention of contributing to the ongoing discourse on theatre in co-communities. The chronological arrangement of the chapters follows, in fact, the course of my professional biography, from my beginnings as a theatre facilitator mainly involved with the disabled elderly and hospitalized children to my work as a researcher, initially investigating my own projects, but then also looking at the projects of many others within various co-communities, thus revealing the development and consolidation of my critical thinking. Consequently, this book focuses on the many ways by which a given co-community utilizes theatre for its own needs (which do not necessarily correspond to those of the commissioning agencies), as well as on the features that make theatre in co-communities a distinctive resistant art form. The bottom-up approach I use here places the co-community at the centre of the discussion.

It is necessary at this point to introduce a number of key expressions that provide the principal analytical tools for the discussion of the case studies in the following chapters.

The theatrical event

There has never been a fixed or stable perception of theatre. Definitions of theatre are always socio-culturally determined and are thus consistently open to challenge. Since the avant-garde movements of the 1920s and the late 1960s, definitions have been considerably expanded to accommodate new forms and practices. Intra- and inter-power relations within the discipline and between related areas such as education and therapy continue to fuel the ongoing debate over the meaning of 'theatre'. Theatre scholars tend to describe this process as a contest between the restricted notion of theatre as the artistic and professional production of a play and the expanded version of theatre as a medium applicable to different settings and goals.[2] This book advocates a more comprehensive approach which views these notions from a Wittgensteinian perspective:

> [...] we see a complicated network of similarities overlapping and criss-crossing; sometimes overall similarities, sometimes similarities of detail [...]. I can think of no better expression to characterize these similarities than 'family resemblances': for the various resemblances between members of a family [...]
>
> (Wittgenstein, 1958, p. 31)

In this respect, it is not only games, as indicated by Wittgenstein, but also theatre that 'forms a family' (ibid.) Moreover, no matter how one defines theatre, it is always manifested as and realized through an event. The idea of theatre as an event, especially as conceptualized by Richard Schechner (1969, 1971, 1982),[3] provides an all-inclusive analytical tool for discussing a range of symbolic activities from playing to performance or theatre production. Schechner clearly stipulates that theatre does not necessarily entail two differentiated groups of people – actors and audience – but two complementary functions – acting and spectating (Schechner, 1966, p. 27). As such, the playing of the elderly within a day-care or rehabilitation environment (Chapters 1–3), the participatory theatre of hospitalized children and student actors (Chapter 4) and the community performances of the Mizrahim, the young Jewish-Ethiopians and the Israeli-Palestinians (Chapters 5–8), are all theatrical events continually challenging the definitions of 'theatre'.

The concept of the theatrical event also helps to reveal 'family resemblances' between the various case studies. All are non-institutional, non-professional, actual, local articulations based on the specific life materials of the co-community. As such, theatre in co-communities belongs in general to 'believed-in' theatre and juxtaposes documentary theatre in particular. 'Believed-in' theatre is an umbrella expression coined by Schechner (1997) to identify those forms of theatre that are in one way or another versions of documentary theatre and predominantly 'explore and express feelings and ideas at the precise point where the personal, the historical and the political intersect' (80). Believed-in theatre seeks to transgress 'pretend' theatre and prefers to occupy the dizzying liminality between realities and fictions, not quite really real, not quite really 'fiction' (Schechner, 1997, p. 99).

Theatre in co-communities and documentary theatre are both theatre arts on the periphery, interpreting 'facts' within the fictive realm of the stage. Both aim to come as close as possible to actual events and experiences by incorporating primary sources through a process of selection, editing and structuring. In contemporary culture, which is full of documented reality, the source materials used by documentary theatre are far more diverse than the historical written documents used in earlier years, when documentary theatre was closely related to the historical drama (Dawson, 1999). Thus the resemblance between theatre in co-communities and documentary theatre is accentuated when taking into consideration the interview-based oral projects of documentary theatre, such as those of Anna Deavere Smith, Jessica Blank and Erik Jensen, Victoria Brittain and Gillian Slovo, and the theatre of the Israeli artist Nola Chilton (see Chapter 6). This oral source-based documentary theatre tends, as Emily Mann (1994) indicates, to work from life – and it is very personal. It challenges public history with living history (Favorini, 1994) and often reaches the stage of therapeutic drama (Lindenberger, 1975). Likewise, it attempts to install excluded voices and experiences and recognizes the power of performance to challenge master narratives and discourses of history (Ben-Zvi, 2007). Nevertheless, the singularity of theatre in co-communities in comparison to these projects of documentary theatre lies in its intensified believed-in quality, which is stimulated by the absence of distancing processes between performer and character, audience and participants, the personal and the political, and the symbolic and concrete/authentic. The core of theatre in the co-community lies in its self-representational tactics, in the ways in which local non-professional performers enact their own stories and perform them in front of their own community, which shares with them the essence of these narratives.

Problem

The theatrical event is based on conviction, sincerity, direct address and local testimony, thus striving to be received first and foremost as 'truthful' and only secondarily as 'theatrical'.

The concept of the theatrical event in its capacity as an analytical tool broadens the scope of the research to include not only the 'dramatic text' (the public performance) but the social and creative processes of the whole 'performance text' as well. This enlarged concept encompasses three interactive events: the gathering (proto-performance: training, workshop, rehearsal), the performing (performance: warm-up, public performance, cool-down), and the dispersing (aftermath: critical responses, archives, memories) (Schechner, 1977, 2002). This comprehensive process-oriented approach is valuable and helpful for the study of theatre in co-communities in general and for the case studies presented here in particular. The 1998 public performance *Here? Now? To Love?* (discussed in Chapter 8), produced by a non-professional group of Israeli-Palestinians and Jews in the cultural centre of a mixed Jewish-Arab town in Israel, could easily be interpreted as a melodramatic love story when focusing on the public performance alone. However, the significant material for grasping the complexity of the performance becomes apparent when the entire creative process becomes the object of observation – the personal stories and experiences contributed by the participants, the conflictual conversations, the moments of silence, the complicated relationship between the Jewish and Arab facilitators, and the arguments regarding the writing and directing of the play. The proto-performance events help to locate the muted gaps in the public performance as well as all the absent explosive materials that the performers managed to 'forget' in order to put the show together. An analysis of the whole performance text, including the responses of the mixed local audience, reveals the inevitable conjunction between the macro Israeli socio-political context, the micro context of the town and the micro text of the theatre group. In this respect, the public performance is not melodramatically escapist but a political realization of the mixed group's ongoing mantra, 'let's put on a show together'.

Articulation/inarticulation

'Articulation' principally refers to the production of speech sounds and is often used interchangeably with the term 'expression'. Hence,

it has become commonplace to see articulation used whenever referring to self-expression through art and theatre. However, it is not so
much in this general sense that I use articulation here, but more in
its political-critical connotation.

Politics in Westernized and capitalistic societies tends to appear in
the form of hegemony. Antonio Gramsci (2004) provides two related
meanings for this term. The first refers to hegemony as a process
whereby the historical bloc – a coalition of social groups – struggles
to organize society around a single, central, dominant project. The
second refers to hegemony as that condition in which a collective
political subject manages to put society in order via a project, which
despite expressing its own central interests, becomes the accepted
world-view. It shapes socio-cultural reality, including its practices,
meanings, values, expectations, beliefs and 'common sense' about
the 'nature' of man and society. Although the hegemony succeeds in
constituting a consensus that is difficult to dismantle, the potential
for the consolidation of counter-hegemony always exists and threatens the power bloc. With resistance embedded in the system itself,
hegemony, thus, is never total. Gramsci, it is important to note,
perceives popular art as a valuable part of culture and as a symbolic
space that may constitute a counter-hegemony. Critical sociologists
after Gramsci have been sceptical about the possible formation of
counter-hegemony within high capitalism, but assume that the potential for isolated, differentiated and small manifestations of resistance
are reasonable. The answer to the question – What defers the rising of
counter-hegemony? – can be found in muted-group theory, a theory
first introduced by feminist anthropologists and sociologists such as
Shirley Ardener (1975) and Cheris Kramarea (1982).

Muted-group theory exposes the power of the hegemonic historical bloc to determine the leading communication system as natural
and right for both dominant and non-dominant groups. This ongoing process renders co-communities inarticulate, which means that
their voices are muted, distorted or under-represented. Theatre in
co-communities demonstrates that the process of silencing is never
total or hermetic. Whenever a given co-community acquires the
opportunity to create its own theatre, it can use the public performance to challenge its inarticulation.

In Israel, theatre projects in, by, and for a given co-community
are usually initiated and commissioned from the top down[4] as part

of state and municipal welfare programmes.[5] The aim of these top-down projects is rehabilitation and social progress within groups with special cultural, social, educational, cognitive and emotional needs. The institutional agencies assume that the marginal group's 'self-expression' through theatre will facilitate or testify to its integration into the dominant order. In contrast, from the bottom-up perspective of the co-community, the theatre project provides a public opportunity to articulate repressed and forbidden life materials that resist or challenge the status quo.

The Other Half (1974), for example, was a performance produced by a group of young dropouts from an underprivileged Mizrahi neighbourhood in Greater Tel Aviv (discussed in Chapter 5) as a platform to articulate their own and their community's perceived second-class status. The title, *The Other Half,* is a condensed critical metaphor articulating the young performers' resistance to being both the oppressed other half of the (national) Ashkenazi community (Jews originating from Euro-American countries) and the inferior other half of the (local) ultra-Orthodox Ashkenazi municipality to which they belong. The local authorities who commissioned the project perceived the performance as a sign of ingratitude, and shortly after the public performance they disbanded the group and dismissed its facilitator/director.

The term 'articulation' has a second denotation indicating a joint connecting two bones. Accordingly, the theory of Ernesto Laclau and Chantal Mouffe (2001) elaborates the second, political, connotation of articulation. In contemporary society, there is a diverse range of various subject positions – such as class, ethnicity, nationality, gender and religion. In order to construct a collective subject and identity, there is a need for a process of articulation between the different subject positions. This connection between the whole and the components is possible because the identities of neither the whole nor its components are stable. Articulation, Laclau and Mouffe suggest, is a political practice that binds various components in a way that affects their identity formation. Thus in a pluralistic society, articulation between the different components that constitute the historical bloc is an outcome of cooperative struggles. But articulation is also crucial in the formation of the counter-hegemonic socio-cultural manifestations of oppression. First, there has to be an exterior discourse that interprets the relationship as oppressive, and then there is a need for articulation between this discourse and the relationship that awakens the social consciousness of the subalterns to their oppressive condition. Only when subalterns perceive their relationship with the hegemony as oppressive do they struggle for change (Laclau and Mouffe, 2001, pp. 18–26). Theatre in the co-community generates a theatrical event that articulates various subjects through its process of creation and consolidates them into a cohesive group.

Facilitating theatre in various groups has made me realize that the powerless are not necessarily aware of their oppressive condition and its social origins. Often they do not even recognize the similarity of experiences they share with others. They have become conditioned to perceiving themselves as poor, irresponsible individuals, and thus as the sole creators of their own condition. The creative process of the self-text motivates individual and group reflexivity, raises social consciousness, and thus articulates the performers as a collective subject with a socio-cultural identity.

To make do with/tactic

The expression, 'to make do with', is borrowed from the theory of the French sociologist Michel de Certeau (1984) and is appropriated here to discuss theatre in co-communities from the perspective of the performers. De Certeau argues that 'common people', 'consumers' or 'dominated groups' 'make do with what they have' (de Certeau, 1984, p. 11). He suggests that ordinary people are not necessarily passive or docile but consume cultural systems and products 'with respect to ends and references foreign to the system they had no choice but to accept' (xiii). For de Certeau, the ways in which dominated groups consume cultural systems are forms of secondary production. Thus, this theory has been adopted mainly by the cultural discourse on popular forms of art such as soap operas, telenovellas, pop and rap music, fashion magazines and movies. In contrast to these private, individual and invisible acts of oppositional reading,[6] I suggest adapting de Certeau's theory to theatre in co-communities, that is, to a site of public, collective primary production offering an empowering effect. The goal of this book is to show how a given co-community 'makes do' with theatre, that is, how it appropriates theatre – a cultural system it did not create – to fulfil its own needs.

The correlation between making do with theatre and articulation is apparent: articulation is the symbolic production of ways to make do with theatre. De Certeau defines these as 'tactics', that is, as non-dominant communication behaviour. In direct contrast to the strategies used by those with power, tactics are the art of the weak; they are bricolage activities by which the powerless group 'takes advantage of opportunities' (37). In the case of dramatic playing with objects, discussed in Chapter 1, the elderly performers, who were not familiar with theatre practices or acting, depended on my instructions. In a simple exercise designed to prepare them for improvisation, I showed them a participant's cane and asked them to name other objects that are similar but not identical to it. They tactically took advantage of this specific means of production with which I provided them, appropriating it for their own articulation needs. One elderly woman, for instance, 'grabbed' the cane and with exaggerated gestures played with it as though she were introducing objects into a large landscape painting. Her tactic of making do with the 'theatre' I had given her at that moment encouraged the other participants to follow suit and, thus, she determined the articulation of the whole scene.

Poaching poetics

De Certeau emphasizes that the tactics of making do comprise a 'poiesis', that is, 'the art of manipulating and enjoying' (xxii), 'poaching in countless ways on the property of others' (xi), or 'playing and foiling the other's game' (11). In this respect, the non-professional performers of the co-communities are 'like nomads poaching their way across fields they did not write' (174). This form of subversive creation is inherently a counter art, a 'nomadic war machine', to use Gilles Deleuze's expression (Deleuze, 1977, p. 148).

The play, *A Plague not Written in the Bible* (2000–02), discussed in Chapter 6, was performed by women from a centre for the prevention and treatment of domestic violence. It presents the daily experiences of women living in violent relationships by focusing on a liberating encounter between three sisters and their mother on Passover eve at the mother's home. The choice of Passover eve is a tactic that enables the performers to play subversively with the central liberation myth of Jewish-Israeli culture; as poetic poachers, the actresses make present the absent feminine of the dominant religious-national narrative. The title of the performance emphasizes the actresses' point of view – battering women is equivalent to one of the ten plagues that God brought down upon Egypt.

Focusing on the subversive tactics by which a given co-community makes do with theatre brings to light the issue of aesthetics in theatre in co-communities. As a local specific form of popular theatre the aesthetics of theatre in co-communities should spring from the given contextual conditions and the local habitus of the co-community. Peter Brook (1968) suggested long since that all forms of popular theatre share a common denominator – a kind of roughness. Popular theatre is a 'rough theatre' which is 'usually distinguished by the absence of what is called style' (74) since it is generated out of and within rough conditions and thus 'anything that comes to hand can be turned into a weapon' (ibid.) But the style that theatre in co-communities lacks is that of the 'synthetic' dominant theatre whose aesthetics, as Jerzy Grotowski (1968) pointed out, is 'super-fluous', depending on rich costume, scenography, lighting, sound effects, music and technology. The rough aesthetics of theatre in co-communities is poor in the sense that it is primarily based on 'the personal and scenic technique of the actor' (15). Moreover, the theatrical events of theatre in co-communities are often 'more social or personal or quasi-religious (ritual-like) than aesthetic. Sincerity and making an honest effort are appreciated. Direct address is stand-ard. At such occasions, "truth" is more highly valued than "beauty"' (Schechner, 1997, pp. 90–1). To conclude, I would say that poaching poetics generates poaching aesthetics, which is in fact a grounded aesthetics. By this I mean a particular aesthetics consolidated within and out of the given circumstances of the specific context of the theatre project. It is an instrumental, activist aesthetics that symbol-izes and embodies a particular, local political necessity in such a way that the political melts into the aesthetic. Grounded aesthetics is thus an actual articulation of the fusion between the political and the aesthetic. This is not to claim that this is the prevalent creative process in theatre in co-communities; it indicates rather that when poaching poetics does reach this stage of grounded aesthetics, the singularity of theatre in co-communities becomes apparent.

The Palestinian youth performance, *Who Killed Achmed Chamed?* (1999) (discussed in Chapter 8), is structured as a court trial depicting the high rate of family 'honour-killing' in the community. Staged outdoors near the ruins of the old Muslim cemetery, the 'White Mosque' and the minaret, the performers' feet on the bare ground produced clouds of dust that visually and physically expressed the violent atmosphere of the impoverished Arab neighbourhood of Ramla. This specific form of grounded aesthetics created a stage metaphor that articulated two interwoven messages: the already articulated social agenda of the youngsters seeking to bring about changes within the traditional power relations in their own community; and the non-articulated historical and social underpinnings of the play which suggested the glorious days of ancient Muslim Ramla and the complex identity construction of Palestinian youth in Israeli society.

These key expressions comprise the theoretical frame within which I shall present the following case studies from a bottom-up perspective. Additional analytical tools and contextual background will be introduced when necessary for each case study.

Theatre in co-communities, as this book seeks to demonstrate, generates empowerment. The empowering process interfaces theatre with welfare – two different 'fields' involved in theatre in co-communities which approach the theatre project from two different ideological perspectives. The welfare bodies initiate and finance the project from a liberal, 'good-will' perspective that represents the dominant world-view, and the social workers in most cases assume that the aim of the project is to integrate, rehabilitate, educate and cultivate the co-community. Theatre practitioners, on the other hand, generally approach the project from a more radical, left-wing point of view that encourages the co-community to engage with theatre as a consciousness-raiser and symbolic weapon with which to stimulate social change. Consequently, theatre in co-communities frequently operates within a contested space and implicitly or explicitly reflects the different, colliding, regimes of power.

The empowering process also interfaces therapy with politics. Modern compartmentalized thinking has tended to split therapy and politics into two well-differentiated domains, although theatre practitioners have always experienced them as interacting socio-personal activities. In this book, I present the formation of the empowering process in various co-communities as the key to developing a political therapy, and vice versa, with the basic assumption that within a community that is powerless in one way or another, the personal experience is always linked to the oppression from which that particular co-community suffers. The theatrical events that co-communities produce constitute significant cultural texts not only to the particular co-community but to the dominant society as well. They contribute the gaze of the co-communities, which as 'outsider-insiders' look at the social structures from a more complex and sometimes alternative perspective. This subversive potential that theatre in co-communities always contains, as a mode of playing in general and as the playing of the co-community in particular, threatens the status quo and fuels the ambivalent position that those with social power hold regarding theatre in co-communities. This book subscribes to the idea that every society needs an apparatus for reflexivity – such as theatre in co-communities – in order to become a more benevolent society.

1
Dramatic Playing as a Tactic for Confronting the Mask of Ageing

The pulse of theatre is sometimes felt in places remote from the centres of the theatre industry. One such site was the geriatric rehabilitation day-care centre where I facilitated theatre in the early 1980s. Before discussing the theatrical events created by the elderly performers and their ways of making do with theatre, it is worth looking at the cultural context of the aged in Western society.

The elderly as a co-culture

The socio-cultural position of the old as other is unique. As the statistical presence of the elderly in Western societies rises, their symbolic annihilation becomes increasingly prevalent. In youth-oriented cultures, old age is perceived as the incommunicable, inaudible, invisible and irreversible other. With 'colonized old age in modern times burdening society with the same weight as colonized native populations used to', the third age, Jean Baudrillard claims, is equivalent to a third-world state (Baudrillard, [1976]1993, p. 163). Indeed, their situation is even more oppressive than the latter, he adds, because robbed of symbolic value, they are an 'asocial slice of life' and 'cultural residuals' (163). 'Anthropologists', says Haim Hazan, 'show favoritism toward the Third World and ignore the Third Age' (Hazan, 2009, p. 60). 'Studying the old-as-other', he continues, 'reveals two types of alterity: that which is *culturally constructed* as different versus that which is *essentially* different' (61). The first is negotiable and changeable; the second is permanent (Archer, 2000). The elderly, thus, constitute an essential co-culture trapped in a limbo state, both included in and excluded from society (Hazan, 1980).

Although consumer culture and the postmodern version of the life-cycle can invent 'the new middle age as a prolongation of the phase of adult life', those with few resources and particularly those confined to geriatric institutions cannot 'facilitate the choice of a "positive" old age' (Featherstone and Hepworth, 1991, p. 374). Moreover, even elderly members of society with resources 'have no sufficient scope to express their personal feelings as distinct from stereotyped responses' (382). Thus, most elderly are fixed and stereo-typed 'in the roles which do not do justice to the richness of their individual experiences and multi-facets of their personalities' (382).

The negative, hegemonic approach to old age described above generates cultural symbols that reinforce the social death and inar-ticulation of the elderly. One of the most salient images of old age, as Featherstone and Hepworth have shown, is the mask of ageing. Functioning as a metaphor for the entire elderly experience, the mask is a pathological manifestation foregrounding the exterior appearance of old people while concealing their inner identity. The mask also defines their basic behaviour, veiling genuine thoughts and revealing only words and actions considered safe enough to utter without incurring repercussions, such as being locked away (374). Primarily, however, the mask of ageing points the way to the death mask, which is impossible to remove (382).

The case studies in this and the following two chapters dem-onstrate how a co-community of elderly performers in a day-care centre makes do with the mask of theatre in order to confront the mask of ageing. The theatrical events created by the elderly are generally in the mode of what I call 'theatre in a closed circle' (Lev-Aladgem, 1995, 1996/7). In such a theatrical event there is no separation between the proto-performance and the performance; rather, it is a single creative process which generates a workshop as performance, based on an actual, immediate and intimate encoun-ter. Such a form of theatrical event enables this co-community to transform themselves from passive patients into active initiators as well as creators of an alternative, symbolic reality. Using the double reference of the theatrical event as a protective shelter (this is not-real) as well as a subversive tactic (this is not-not-real), the elderly performers criticize their status in society in general and in the day-care centre in particular without exposing themselves to repercussions.

The setting of the theatrical event

The day-care centre, a beautiful new building surrounded by trees and gardens, is located in a respectable suburb of Greater Tel Aviv. The rationale of this geographic strategy was humanitarian: to disengage the elderly from the hospital environment following prolonged hospitalization and to reintegrate them into society by supplying rehabilitation services within the community. Despite these good intentions, the local citizens, believing institutional activities connected with disability might corrupt the image of their wealthy private suburb, tried to prevent the construction of the centre and later ignored it, as though the place and its elderly members were completely invisible. Ageing is a natural, universal phenomenon but the reception of its various manifestations is culturally determined. Modern Western society is built upon a definite split between two categories: those who are aged and those who are not. Because the social condition of the aged is perceived as non-acceptable, their image becomes pathological, and ageing is perceived as a cultural disease (Hazan, 1988). Thus the positive geographical ideology appeared insufficient to overcome the double 'dowry' that these elderly people brought with them: defined as medically *and* culturally diseased, the physical incapacity of the residents, who suffered from heart disease, strokes, diabetes, moderate dementia and other medical conditions of the aged, was exacerbated by social exclusion from the community.

Each morning, the elderly clients were transported by special vehicles to the entrance of the building, where the waiting staff helped them inside with wheelchairs or walking canes. The same routine was repeated in the afternoon when they were taken back home. All the activities behind the closed doors were of a clearly medical nature, mostly physiotherapy and occupational therapy. The elderly clients, sitting in fixed places in the large hall, were usually silent, cooperative and obedient as they quietly concentrated on their work or food and waited for their treatment or help to the toilet. The staff, preoccupied with fulfilling their bureaucratic and technical obligations, were usually too busy with the professional aspects of their work to relate to the day residents as individuals. The building had large curtained windows, and peering through them gave the impression of a slow-motion silent movie. This image seemed to echo the general anthropological accounts of the aged in contemporary society: being unable to contribute to the rapid technological pace of the modern world, they are doomed to live almost invisibly in a socially, mentally and physically shrinking world, in which their life perceptions no longer match their distorted existence in time and space.[1] As a theatre facilitator, I soon realized that the physically and/or mentally challenged members of the centre would not be able to memorize long texts, or perform complicated *mise-en-scènes*. I thought it worthwhile, therefore, to offer them theatre as an event within the framework of playing that would suit their preserved life resources.

I set out with the notion that dramatic playing could act as a path to articulation, enabling these elderly people to interrogate their socio-cultural and existential being in its particular manifestation in the centre. Consequently, my focus was on the potential of the theatrical event to encourage the elderly performers to create an alternative reality by taking up various fictional, imagined roles as a means to undergoing symbolic change. In this respect, I also looked for a correlation between the struggle for meaning through articulation and the process of emotional and mental self-empowerment.

Dramatic playing as a theatrical event

Dramatic playing is generally defined as 'the free play of very young children, in which they explore their universe, imitating the actions and character traits of those around them' (McCaslin, 1990, p. 4). With children, dramatic playing is more of a training ground, a safe universe that enables them to prepare themselves for the future. The functional, telic orientation of dramatic playing is rational and clear. By taking the role of 'significant others', the child acquires the language, codes and symbols of society and in the process becomes a significant other. With the old, and especially with the disabled elderly, who are socio-culturally reconstructed as non-significant others, the perception of dramatic playing should be different. It should stem from an approach to playing not as an instrumental activity with practical ends but as an end in itself.[2] This autotelic event can provide the adult player with an 'optimal experience' of being totally and intensively absorbed in the symbolic activity (Csikszentmihalyi, 1990). This optimal event enables the elderly co-community to make do with playing for their own needs by making something happen symbolically that otherwise could not happen at all.

This event is potentially therapeutic and political at the same time, because dramatic (symbolic) activity has the power to generate both individual and group empowerment. Moreover, dramatic activity is not merely a reflection of a given reality, but an articulation that contributes to the dynamization and consolidation of self and society. 'Self and society are generated as they are expressed', suggests anthropologist Edward Bruner: 'every expression is also a change. Culture changes as it is enacted, in practice' (Bruner, 1984, pp. 9–10). I would like to posit an alternative, wider definition, one that includes dramatic playing as a theatrical event generating a lifelong experience. Dramatic playing with the disabled elderly is a foregrounded activity; performers ostentatiously use an object, not according to its literal usage but as a means to creating an alternative place and/or impersonating somebody other than themselves in front of an observer who is able to distinguish between this activity and every-day behaviour.

From individual behaviour to group formation

Most of the members of the centre passed the time in occupational therapy, habitually shutting themselves off from their surroundings as they sat with bent head and fixed gaze on their almost automatic hand movements. Did they find in such activity a partial escape from boredom and a degree of satisfaction at still being able to do something, or was this a sign of loneliness, social disengagement, and an increased withdrawal into the self? (Kastenbaum, 1981). Suffering from various disabilities, they were almost totally immobilized and rarely talked to each other. Did they still retain the basic need for contact and communication? Could they articulate it through gesture and voice despite their limited skills? Dramatic playing in a small group seemed a suitable setting for generating a creative encounter that might re-establish direct, personal and face-to-face communication. Each week the centre accepted one or two new members while the older and least well residents gradually disappeared into hospitals or dedicated nursing homes. Thus, I organized an open group,[3] which could welcome new participants and also ease the parting from those who had to leave.

While the elderly attendees at the centre were outwardly coopera-tive with and loyal to the staff, in our private talks I sensed that this was simply a mask of docility and that they were really concealing thoughts, emotions, opinions and stories that they believed should not be expressed openly. A dramatic encounter might provide them with a neutral, free zone where they could express whatever they wanted without harming themselves or the staff. Playing is Janus-faced by nature. It is 'a stepping out of "real" life into a temporary sphere of activity with a disposition all of its own' (Huizinga, 1955, p. 8); it is 'separate: circumscribed within limits of space and time' (Caillois, 1961, p. 9). In fact, the relationship between playing and reality is more complicated. Bateson described it as the 'map-territory relation': 'they are both equated and discriminated' (Bateson, 1976, p. 125). This double reference could make dramatic play appealing for the elderly, who might feel safe enough to make statements that in the context of symbolic activity are simultaneously true and false, pretend and real.

The staff, believing the new event would not upset the social status quo in or out of the centre, approved the theatre project. As a young theatre facilitator, a complete stranger and an outsider, I had first to gain the confidence and trust of the day residents, and in the process become an 'organic intellectual' within their co-community, organizing and consolidating their individual resistance into a cohe-sive group. I employ here Antonio Gramsci's concept of the organic intellectual who plays a central role in the formation of the counter-hegemony. The organic intellectual helps to motivate, organize and consolidate the resisting group (Gramsci, 2004, pp. 35–52). In this respect organic intellectuals operating in co-communities are those who hold a critical, socio-political stand and choose to commit them-selves to social and political involvement. Seeking a way to assimilate into their community, I visited the centre daily, using two basic tech-niques I had acquired as an actress: acting 'from moment to moment' and from the 'here and now' of the field, I tried to respond spontane-ously and intuitively to every sign they expressed. Imagining them as protagonists of a slowly emerging narrative, I listened and observed, taking into consideration only the immediate 'given circumstances' of the actual situation and trying to reject previous generalizations I might have held about old age and the handicapped. In this initial stage, I did not tell them that I facilitated theatre, but that I was a volunteer helper serving meals, helping them reach the toilet and otherwise giving them attention whenever needed.

Because the staff – comparatively much younger than those attending the centre – had over time taken on the serious, sad mask of their clients, the residents not only welcomed my cheerful face, humorous tone and expressive body language but were also both amused and curious about what was going to happen next, as though they sensed that there was a more serious reason for my visits. They often repeated comments such as: 'What has a young girl like you to look for here?' (Esther). 'It's nice having you here, but it is not a place for you' (Golda). 'You are always smiling, that's wonderful, but looking around you at these faces can only depress a person' (Menachem). 'In a few days you will disappear' (Ben-David, a new widower).

As a way to prepare them for the improvisations, I tried to establish a personal relationship that would not be forced or pretended but constructed on the grounds of mutual interest and appreciation. While helping them with their immediate needs, or when finding individuals on their own, I talked to them casually, not asking too many questions, but trying to gain information indirectly by giving them the space and time to respond according to their personal timing. The efficacy of a communicative process depends on the adjustment of different rhythms, and one must take into consideration the slower tempo of the aged.[4] I began to perceive the elderly people as potential friends able to ask me questions about my life, work and feelings as well. The initial process was completed within two months. My acceptance into their community was first manifested in the reception ceremony performed each morning. Some waved to me slightly, some nodded, while others winked or smiled. These little gestures, almost invisible to a non-observant eye, clearly signalled their satisfaction at seeing me again. At this stage, they would make the following comments during our private talks: 'I was waiting for you, you aren't going to neglect me, are you?' (Ben-David). 'I talk to you like to a daughter, except my daughter is always so busy that she doesn't really listen to me' (Golda). 'I can't talk like this to Lea or Sara [staff members], they are good, but I don't think they really have the patience for us, they are fed up with our troubles' (Israel).

'It's fun talking to you, better than to Jacob [the psychologist]; he's quite arrogant, you know, and I don't think he has any justification for it, of course this is between you and me only...' (Chaya).

This was a good time to bring up the idea of theatre, adjusting the formulation of the offer to each person according to the manner of our conversations. To Chaya, for example, I said that I had got to know some people who seemed to be very interesting and humorous like her, and I thought that as a group we might have more fun together. I told Ben-David that he shouldn't restrict himself to making friends with me only, and that I would like to include him in my working group so that the others might enjoy his storytelling as well. To Menachem I said that participating in a group activity might provide a fresh change to his preoccupation with being angry and unhappy about his physical situation. To Israel I presented my plan more directly. He loved the theatre and was always happy to tell me about the performances that he had seen back in Poland years ago before the Holocaust. He recalled stories about the great actors in those days, how he himself had loved to act, and as a youngster had dreamt of becoming an actor. It was obvious he would like to join the group and become a central participant.

Surprisingly, the idea of dramatic playing did not seem strange to them. However, they clearly expressed a lack of confidence in their ability to do something meaningful and to become significant again. Similarly, Claire Michaels, in her article 'Geriadrama', reports that 'frequently, the elderly underestimate their ability to do anything worthwhile, especially after arriving at a nursing home or rehabilitative institution' (Michaels, 1981, p. 175). One of the reasons for such a low self-image is what I call the 'stereotype circuit'. In the course of a lifetime, people assimilate various stereotypes that become 'common sense' and fixed cultural coordinates. Old people therefore tend to perceive themselves according to the prevailing negative images of old age, and fail to acquire different, more positive substitutes. It took several personal encounters devoted to encouraging the day-centre clients to interrogate these stereotypes before they came to believe that my interest in their lives was genuine. Their final acceptance was either cautious or pessimistic: 'O.K., I will do it for you', or 'Well, O.K., what do I have to lose anyway?'

In order to give each participant enough attention and space for articulation, I organized two groups of 6–8 members. My main objective was to construct the theatrical event so that it would develop organically in correspondence with the participants' capacities and that it would demand minimum physical energy while facilitating maximum articulation.

Structuring the event

The formative principle of the encounter was based on the idea that among adults, and especially among the disabled elderly, dramatic playing should start with an 'exterior' activity. It is cognitively easier to perform a foregrounded action while the focus is on something 'out there' that represents the protagonist. Playing with objects and spaces is a simple dramatic activity on its own, but it is also the preparatory stage for the interior, self-centred activity of role-playing. Acting out a character demands a certain readiness to openness and mental exposure, which might be threatening for amateur participants at the beginning of the event, especially the disabled elderly. Consequently, I structured the event to progress gradually from very simple to more complicated dramatic playing; from playing with objects to playing in/with imagined locations, and finally to acting out a role or character. The principle underlying this technique is drawn from two different sources.

Stanislavski's *An Actor Prepares* (1989) develops a training method that moves from exterior-focused to interior-focused activities: first his students are sent out to observe their surroundings using all their senses in order to memorize as many details as possible. At a later stage, the students are requested to bring a specific photograph to life by adding various objects to it, imagining the location and space, and then introducing themselves into the story. Only at the third stage is the focus of activity moved to the self. Working with a partner, students concentrate on their own as well as the other's body movements, voice, reactions, feelings and thoughts, preparing themselves to create a dialogue with a given character. In *The Play's the Thing*, Marina Jenkyns expands on Peter Slade's distinction between projected play and personal play out of which dramatic playing evolves: 'Projected play, as its name suggests, is concerned with the child focusing on the tools of the play activity [...]. Personal play involves the activity within the self [...]. These two forms of play I believe develop the ability to be in role' (Jenkyns, 1996, p. 14).

The following three excerpts from my field diary demonstrate the form and content of the elderly performers' theatrical event, from playing with objects to creating places, and finally to acting out a role or character.

Dramatic playing with objects

Participants: **Esther** (74 years old, with a smiling face, short and heavy, diabetic); **Ben-David** (75 years old, new widower, very thin, very wrinkled face, usually in a bad mood, cries easily); **Motterrem** (68 years old, a spinster, tall and thin, suffers from heart disease); **Miriam** (70 years old, religious, with bright eyes, thin, paralysed left side of face and left arm); **Menachem** (75 years old, religious, short and skinny with small cynical eyes, in a wheelchair); and **Golda** (74 years old, straight white hair, short with a bent back, a limp, and semi-closed left eye).

I asked Esther to give me her walking cane and placed it in the middle of the circle. Esther's wooden cane is hand made, with engraved wavy lines, black trees and a deer in the middle. Our first exercise was to observe it carefully and then describe it in detail. Esther gradually became amused and quite flattered at seeing one of her belongings, 'especially the one which I am not so much in love with, as you know, become such a celebrity...'. After everyone had finished reporting their own observations of the cane, Esther said: 'I have never really given it a good look, a cane is a cane. Well, now actually I begin to think that it is quite a beautiful piece of work.' I then asked the participants to think about all kinds of other objects that are similar but not identical to the cane.

Motterrem grabbed the cane and pointed excitedly with exaggerated hand gestures: 'Here is the sea, look! In the middle there is the land with trees and roads and animals and a big zoo, and here at the top we have a fish, do you see this?'

Esther said it reminded her of a hammer. I asked her to use the 'hammer'. She laughed at my request and then took the 'hammer' and 'nailed up' Golda's back with some 'nails'.

Ben-David insisted that it was an ax, and to convince us he started to 'chop' some 'wood' with his 'ax'.

Miriam waited calmly for her turn, and then demonstrated cheerfully: 'Here is the post office antelope, and his role is to carry Estee [Esther's nickname] quickly, quickly, wherever she wishes!'

Golda, who said at the beginning of the encounter that she had come 'just to watch', soon found herself absorbed in the play: 'It is the Independence Day plastic toy hammer', she cried out. Then she took 'it' up and started to hit my head to the laughter of all the others.

At that moment Menachem wheeled himself swiftly into the room. Usually those participants who entered late did so very quietly in order not to disturb the activity. Menachem, however, wheeled his chair into the middle of the circle and said: 'I would like to say something about the play we performed last week. I was thinking about it. You know I am religious. I asked myself whether it is proper to play according to the Jewish religion. Well, the answer is positive! You see, I found out that according to our wise men, the world was created for our enjoyment, and he who does not enjoy the world is a sinner!' The group reacted enthusiastically to this idea: 'Yes, yes, you are right, you made a good point.' I told Menachem about our present play and he complimented the participants on their bright ideas.

'I did not know that my stick is so important', said Esther to Menachem laughingly. 'Yes, we amused ourselves', followed Ben-David, 'a stick is not just a stick after all', he said with a wink. 'So, was it worthwhile playing?' I asked. 'Certainly', Ester answered. 'It stimulated our imagination, we could suddenly see things we did not see before.' The word 'imagination' lit up Motterrem's face; she grabbed the cane again and played as though she were hiding money in the cane, and then pouring wine secretly into it, as if nobody would ever know.

A walking cane is one of the most dominant symbols of old age. Playing dramatically with it provides the participants with the opportunity to transform it, to create new images out of it, to master it, to laugh at it and thus to perceive it in a more friendly light. Even a stick can provide a stimulus for creative thinking and articulation by altering the player's spirit and frame of mind. The above example demonstrates the transition from passive action (observing and reporting) to a more active involvement (describing and using a symbolic object). My goal was to guide the dramatic playing as if it were a natural and familiar series of actions. On the one hand, it matched the participants' level of energy; on the other hand, it also made them more vital and alert. The process introduced a pleasant atmosphere in which the participants were able to develop their creativity, and eventually they made do with the 'theatre' I gave them in their own ways and according to their own needs. Adult dramatic playing also has a unique dimension, that of the reflexive, responsive phase, which should be considered as an integral part of the event. The facilitator enables and encourages the performers to respond spontaneously and openly to their playful actions, to think them over, to make sense of them, and thus to reintegrate their significance into their present-day lives.

Dramatic playing as the creation of imaginative places

Participants: **Israel** (73 years old, in a wheelchair, with bright eyes and ruddy cheeks); **Chaya** (69 years old, very open and friendly, heavy, with one arm and leg paralysed); **Dina** (84 years old, with a pleasant smiling face, diabetic); **Ben-David**; **Menachem**; **Eva** (82 years old, with a rough voice and very heavy body, in a wheelchair).

The performers gathered enthusiastically in our activity room, commenting that they became happy just by thinking 'about our family meeting'. It was very cold this morning. In order to get warmer, we held hands and shouted three times: 'Good luck, good luck, good luck!'

I said: 'If I could take you to any place, where would you prefer to be now?'

Israel asked to play first. He told us that he was now in Warsaw, sitting in a theatre hall. On the stage the famous actress Esther Rachel Kaminska was performing the lead role. He was inviting us in (verbally and also with a hand gesture). We asked him questions about the plot of the play and the life of the actress and he tried his best to satisfy our curiosity.

Chaya was next. She wanted to be healthier and go back to work in her grocery store. She directed us into her shop, indicating all the shelves and goods, and asking each of us in a coquettish way: 'What would you like to buy?' 'Can I help you, dear?' or

'Here is a little boy coming in', pointing to Israel and patting his head. 'Here, take some candy.'

Dina was sad and apparently unable to join our magic journey. 'Take me to the hospital', she requested, coughing. I asked: 'Is that where you wish to take us on this extraordinary occasion?!' She fell silent. We all waited patiently. In a very low voice she began to speak: 'I am now in Venice. Look at all these pigeons down here on the plaza. Can you see over there all those beautiful trees? Look up high and see the clear, blue sky' (she pointed toward the window at the gray, cloudy sky). Her tone grew louder as she took us to the countryside, showing us all the little houses with the red roofs and the glass workshops. 'Here are the snowy Alps. Look down. We are approaching Switzerland, so warm and nice here.' She no longer coughed, but smiled in high spirits. She would have liked to go on and on...

Eva said very strongly that she would like to stay here where she felt well. To make us understand her point, she said: 'Come with me to the Russian commune!' She demonstrated to us with her hands how hard she was forced to work, and told us about her son whom she had just lost. She was sad for a while, then began to cheer up: 'I am a young woman in a citrus plantation in Israel, in a small village, Zichron-Yaakov. I am working hard but I am happy.' To convince us, she forgot her disabilities for a moment, bent down and picked up 'heavy orange boxes'.

Ben-David took us to the youth club where he used to teach carpentry. He described the place in detail and then demonstrated how he used to teach. He approached an imaginary girl: 'You can have the hammer that you made as a present, if you wish', he said in a fatherly tone.

Menachem wanted to be on the beach and chose to take me there with him. The performers laughed and said they would follow us but promised not to 'disturb us'. Menachem looked around, describing the height and colour of the water. He showed us two young girls who were sunbathing and a guy who was selling ice cream. I became 'annoyed' that he was neglecting me and did not buy me an ice cream. 'Just ask', he said, 'I will even buy you two.' Then he looked at his watch: 'Oh, it is four o'clock, I must hurry home.' Chaya and Eva began to laugh: 'Let us see what Menachem will tell his wife?'

They immediately began to clean 'the sand' from his shoes and pants.

Menachem announced: 'I am at home now, hello.' He approached me. 'Where have you been?' I reacted hysterically. 'I was so worried!' Menachem did not lose control, but very calmly said: 'I was at a secret meeting with the underground. You know we are planning to sabotage the English occupier. But you are right, next time I will find a way to inform you.' Everybody laughed and Chaya said: 'Menachem not only deceived his wife but also succeeded in hiding it from her!' Then Menachem looked at me and said: 'You won't tell my wife where we were...?'

I asked the performers to respond to our playing. Eva said that it had allowed her imagination free rein and Dina noted: 'Yes, with imagination we can see and do so many things. It was as if it were real, it really was like a journey.'

The psychological effect of playing triggered by the theatrical event was significant. Absorbed in play, Dina stopped coughing and her mood radically changed. Eva, the oldest and weakest member of the group, performed physical movements that her physiotherapist would never have believed. Although this was a temporary change, it was meaningful as a symbolic victory over human limitations. Moreover, transforming the playroom into various places gave the elderly the opportunity to deconstruct their past and memories. By living through a part of the past as present, they created a new experience, and this moulding of past into present engendered an atmosphere of excitement and celebration. This was facilitated through minimal interference, directing the activity from 'moment to moment', from the 'here and now' of the event and, when necessary for the effect flow, getting into play and taking up roles. Finally, through the event the performers implicitly articulated their innermost feelings about being disabled, dependent and immobilized. The theatrical event initiated a more fruitful approach than simply discussing these issues directly. Performing the past in the present enabled each individual to challenge their self-image and to negotiate with it in front of their friends. It was also fun for them because they were well aware of the boundaries between reality and illusion, the possible and the impossible. Nevertheless, they made do with theatre and created a moment of impossibility. They were active, they were imaginative, they responded to one another, they had fun, and they took the situation into their own hands, presenting themselves once more as significant others.

Dramatic playing as acting roles/characters

The following event took place on Tu b'Shvat, the Jewish festival of the trees.

I asked the participants (Dina, Motterrem, Ben-David, Israel, Miriam, Chaya and Eva) to look outside and observe the trees. Then I asked them to imagine other trees that were not in the garden. After a while I instructed them to choose a tree and present themselves as the chosen tree.

Dina: (ramrod straight) I am a birch tree, I am white, tall and beautiful. They collect my sap into a cup and drink to satiety (she pantomimed the action).

Motterrem: I am a pine tree. I am tall with beautiful leaves. I have a wide, strong trunk (she indicated her spread knees). Sometimes blood flows from me...

Ben-David: Blood?

Motterrem: Yes, blood, flowing downwards (she illustrates), as it does among women each month.

Ben-David: I am an olive tree. I am always green. I live for a thousand years!

Israel: I am an apple tree.

Eva: I am an oak tree.

Chaya: (declamatory) I am a cedar tree!

I asked the birch tree to meet the pine tree. At first Dina refused, saying that she did not know what to do, but she quickly found her way.

Dina: ('pricking' Motterrem's shoulder with her fingernails, which represented the prickly branches of the birch tree in the game; Motterrem's shoulder was a pine branch.) What is this red stuff coming out of you? Do you drink it?

Mottereem: No.

Dina: Do you paint with it?

Motterrem: No.

Dina: Do you cook with it?

Mottereem: No!

Dina: So what is it?

Motterrem: It's from God. You don't use it for anything!

Dina: What sort of thing is it? My sap is white, sweet, very tasty...

Motterrem: And my leaves are bigger and more beautiful than yours. And here, beneath my leaves there are people sitting (she beckons the other participants to come to 'rest in her shade'.)

The second encounter was between the olive tree and the apple tree.

Ben-David: I will live a thousand years.

Israel: That's right, but I am much younger than you!

Ben-David: Who are you? I don't know you at all!

Israel: I am an apple tree.

Ben-David: Oh, you probably need a lot of care.

Israel: No, not so much, some water, fertilizers and trimming.

Ben-David: Look at me, I am always green. My oil is the best!

Israel: Well, I prefer corn oil...

Ben-David: Nobody uproots an olive tree. I am a holy tree!

Israel: They don't uproot me either.

Ben-David: They can with a licence.

Israel: Look, I agree that you are stronger than me, greener and more important than me.

Chaya: Why do you agree? You have wonderful fruit and it is healthy. You have a lovely smell. I love apples!

The next encounter was between the cedar and the oak.

Chaya: I am tall, very tall (demonstrating with her hands), and I am a beauty (putting on a seductive face).

Eva: So am I.

Chaya: I am a historic tree! It was from me that they built the Temple!

Eva: But you are so expensive! I am cheaper.

Ben-David: (helping out the oak) and quite strong.

Eva: That's right. I am strong and quite beautiful!

Chaya: I am the real beauty! I live a little in Israel but mainly in Lebanon. I am famous. I am a symbol!

The participants were very happy with their performance. Chaya said:

'First, I have learned about trees things I didn't know before, and besides, today it was like a real show, we were like real actors!'

In this event, the performers acted and reacted to each other through their own reflexive perspectives as old people. The self-awareness and reflexive viewpoint are what endow the 'tree game' with its real internal meaning. As trees, the performers could articulate their inner selves, their problems, wishes and world perception. Dina, frail and stooped, could straighten up in the game. She chose to be a tree that is not only beautiful, but also beneficial to its environment. Out of play Motterrem was always proud of herself as an ex-nurse, devoting her life to the public. In play she expressed complex feelings about her femininity and her need for relationships and intimacy. In a way, all the participants played an inverse reality. These semi-symbolic inversions expressed what they felt about old age, but also enabled them to change it, if only temporarily. Nevertheless, this temporal micro-universe affected them deeply because although it was only playing, it was also really happening and, moreover, in front of witnesses.

With adults then, dramatic playing goes beyond its developmental function as a tactic for achieving mental and social maturity and becomes instead a tactic for group and self-exploration. Dramatic playing can make present that which otherwise would not be present and thus it is not merely a reflection of reality but an alternative construction parallel to reality. For the elderly co-community, especially those attending a geriatric day-centre, dramatic playing can be a suitable symbolic resource for transforming individuals who passively 'receive' projects into active initiators and creators of an alternative reality. The playing process enables the elderly players to retain their self-respect, to do something meaningful, and also to entertain themselves. They receive attention and appreciation for their creative contributions, they empower themselves and begin to feel like 'normal' human beings once again.

In their playing the elderly acted as poachers presenting their own 'art of manipulating and enjoying, escaping without leaving' (de Certeau, 1984, p. xxii). This was manifested in several enduring signs. The performers continued their playing behaviour even after the end of the dramatic encounter. Back in the main hall, they addressed each other by the names of the roles they had played. They sometimes performed spontaneous short improvisations in order to deliver an actual disguised message concerning another inhabitant or staff member. In time the players adopted friendly contacts with each other, spending time together, especially before and after meals. Menachem and Ben-David, who discovered that they had both been active as youngsters in underground movements, performed a survey to uncover additional ex-fighters, and then held talks discussing and arguing over what they considered to have been the true historical facts. The performers sometimes demanded that I or the staff treat them as if they were really kings, rulers, fighters, beautiful young ladies and successful people. Although at times they tried to push their chosen alternative roles beyond the formal frame of the activity, they always did so playfully, using the double reference of play as a protective shelter. The play of double meaning outside the playroom became one of the main tactics for liberating their voices and articulating their truth without annoying the authorities.

The elderly performers thus made do with theatre for their own needs. They put on the mask of theatre in order to put down the mask of ageing. 'It is far better to be an actor', said Menachem, 'especially a leading actor, than an old man.'

2
Performing the Scroll of Esther: Articulating Power through Symbolic Inversion

In the previous chapter I presented the forms of dramatic playing articulated by the elderly performers of a rehabilitation day-care centre as tactics by which to subvert their mask of ageing. As I indicated, it was the 'tree game' in particular that produced semi-inversions of the performers' oppressed reality. In this chapter I focus on the form of dramatic playing through which the performers introduced a fuller and stronger symbolic inversion of themselves and the centre.

During the Jewish holiday of Purim I facilitated a humorous theatrical event following the model of the traditional *purimshpil*. The holiday of Purim commemorates the committed faith and astonishing survival of the Jews following a plot against them in ancient Persia, as inscribed in the biblical book of Esther. The *purimshpil* – a parodic playlet performed by amateur players, celebrating the merry Jewish festival of Purim – constituted the highlight of the holiday in the Diaspora. It takes its tone from the Scroll of Esther, a holy scripture which tells the story of the salvation of the Jews from the hands of the evil Haman, but which relies on a structure of symbolic inversion that substitutes spirituality with a carnivalesque and profane atmosphere. In this short book of only ten chapters, God is never mentioned, but 'feast' is repeated twenty times. Wine, food and women constitute an evident theme in the plot and colour it with the hue of folly. Additionally, the king, Ahasuerus, is not characterized as a king proper. He has no authority; even his wife, Vashti, refuses to obey him, thus revealing a spark of genuine feminism. He is indecisive, constantly enthusing over every piece of advice he gets from his troublemaking counsellors.

Disguise is another basic element of the plot. Esther pretends to be a local Persian girl and only reveals her true Jewish identity at the very happy end. Mordecai, her uncle, circles the gates of the palace like a secret agent, seeking evidence of the conspiracy against the Jews and the king. Role reversal is also a dominant theme. Haman, the highest ranking minister, is demoted to the role of servant, while Mordecai, the lowly subject, is raised to a position second only to the king. Vashti, the queen, is sentenced to death, and Esther, the poor orphan, is elevated to the highest rank as the Queen of Persia. The main and most important reversal, of course, is that which takes place between the Persian oppressors and the Jewish oppressed. But, as already stated, this serious core plot – the escape from potential disaster to salvation and joy – is enveloped in a ludicrous frame.

This carnivalesque ambience continues to be manifested when the text of the scroll is used as the basis for the *purimshpil*, which developed as a custom for the festival of Purim in the sixteenth century, and was enacted by poor folk, amateur actors and singers, who gathered in public places and wandered from one wealthy home to the next, performing parodies in exchange for food and a few coins. While these parodies were based mainly on the Scroll of Esther, there were also allusions to other biblical and traditional themes. Performatively, the *purimshpil* was particularly influenced by the German *Fasnacht* (Ziv, 1995), and it flourished as a folk tradition in Europe until the beginning of the twentieth century. In modern-day Israel, especially among the secular Jews, the centrality of the *purimshpil* in the festival of Purim has faded, but the holiday still maintains its inverted nature in the customs of masquerade and feasting. The traditional *purimshpil* seems to have become little more than a memory among those now elderly Jewish women and men who emigrated from Europe before and after the Holocaust.

The geriatric day-care centre as potential place for symbolic inversion

The *purimshpil* I discuss here was created in Israel in 1983 by several elderly members of a day-care centre and differed greatly from the traditional one.

The elderly created their own unique brand of *purimshpil*, although it maintained some typically carnivalesque features. By exploiting the inversion that characterizes Purim and making it a unique occasion for articulation, the elderly performers created an alternative, resistant metaphor for their existential problems as old people. According to Barbara Babcock, 'symbolic inversion may be broadly defined as any act of expressive behavior which inverts, contradicts, abrogates, or in some fashion presents an alternative to commonly held cultural codes, values and norms [...]' (Babcock, 1978, p. 14). Symbolic inversion relates more specifically to the ancient (in)discipline of 'the world turned upside down', which was the main ideological and performative principle of carnival in its heyday in medieval and Renaissance times. The technique refers to an inversion of bipolar opposites, the creation of a second world outside the established reality.[1] This kind of inversion needs a firm normative and ordered social system from which to spring, as Umberto Eco notes: 'Without a valid law to break, carnival is impossible' (Eco et al., 1984, p. 6). In a post-modern secular, pluralistic, multistructural and sometimes anti-structural culture, the world turned upside down principle is practically meaningless on the macro level. But it can still work as an artistic enclave in well-defined, restricted, closed social organizations such as in Goffman's 'total institutions' (Goffman, 1960, 1961), and in semi-closed ones such as the 'greedy institutions' described by Coser (1974). In such places, symbolic inversion has a profound significance beyond its psychological function as a safety valve for repressed emotions. It also offers a socio-cultural tactic by which to experience legitimately a temporary, alternative reality. Symbolic inversion reveals that the power of any social organization is neither natural nor permanent but simply a matter of social agreement.

Day-care centres that fall into the more traditional, medical type are examples of 'greedy institutions'. In contrast to earlier types of old-age 'homes', they are ideologically and politically correct; rather than being merely a holding centre for the aged, they are established to facilitate the daily lives of the elderly and to support and enrich their remaining time. In fact, however, as a result of the severe physiological, mental and social limitations of its clients, the day-care centre tends to operate as a 'greedy institution', holding full authority, supervision and control over the activities of its elderly and dependent clients. The old people usually play the part of cooperative, obedient and satisfied customers in order to obtain the help they need. Because of the immobility of the old people, the day-care centre reinforces their separation from the outside world, and the strict, tightly-structured daily programme of activities contributes to the erection of symbolic boundaries between the elderly and the staff and between the elderly and outside society. Indeed, this institution reinforces the widely-held image of the elderly as roleless, disengaged and culturally invisible.[2] The day-care centre is thus a place in which symbolic inversion can manifest its function and meaning, as the socially peripheral is made symbolically central and as what is accepted as culturally normal comes under question. Symbolic inversion, however, requires a special playful occasion, which in our case was the festival of Purim.

The *purimshpil* event

I presumed that the participants from the day-care centre would be enthusiastic about the idea of a humorous event following the model of the traditional *purimshpil*, as most of them were European immigrants familiar with the tradition and they were usually very happy to recall their past memories and enact them in playing. Some of them did indeed remember taking part in a domestic *purimshpil*; one woman even talked about a 'big show in a large hall'. However, they showed no enthusiasm for re-creating such an event. They sat quietly, staring at me, and waited for me to change the subject. The traditional *purimshpil* seemed to have lost its power and to have become a 'cold empty form' (Moore and Myerhoff, 1977, p. 14). As I shall demonstrate, however, the elderly 'poachers'[3] found a way to manipulate and to make do with the traditional material for their own ends. By using an old model for new purposes they remoulded tradition and thus were able to 'indicate problems which might not otherwise be recognized' (Hobsbawm and Ranger, 1983, p. 12). Tradition can not only be invented but can also become a stimulus for invention; an occasion to 'construct something new out of old cultural materials' (Moore and Myerhoff, 1977, p.10). My field notes, which I directly quote below, read as follows:

'Wouldn't you like to be the king'? I asked Menachem. Menachem, sitting in a wheelchair, holding his stick, gave me a sharp look: 'No, certainly not!' 'Why' I asked. 'Because I don't want to be stupid... I have enough... I don't need to be stupid also', he said grumblingly. Chaya, sitting with her lame leg stretched forward and one of her hands bent, smiled cunningly: 'I don't mind to be Esther, but not that virgin type from the scroll. All my life I worked in a grocery, behind the high desk, I was always nice to the kids... so what do you say Menachemke?' She winked to Menachem. 'No', he insisted, 'I don't want to be stupid, drunk and kill my wife like Ahasuerus!' 'So, maybe you don't have to', whispered Dina. At that moment an idea entered my mind. 'You can fill in the gaps in the story', I said, 'those situations which are missing.' 'That means I can be a clever king?' asked Menachem. 'But that will change the whole traditional story', reacted Eva. 'So what', said Chaya, 'Let's see what can happen, it can be fun.' Then we started to discuss the first improvisation. They asked me to join them and play a role.

This scene, which happened before the drama itself, during the social gathering, is nevertheless an important part of the total performance text, which in Schechner's sense usually encompasses three interactive events: the gathering, the performing and the dispersing.[4] This mode of democratic, spontaneous negotiation was not a habitual activity in this co-community. The elderly clients lived by the unwritten law: 'Be nice and obedient to the staff, then you'll get your physiotherapy on time, somebody will help you out to the toilet and you will be served with your favourite food.' Here, in the theatrical event organized as a separate, closed and intimate frame of play activity, they felt more freedom. As the theatre facilitator, they saw me as a middleman. On the one hand, I was a part of the staff, an authority figure; on the other hand, I had established a unique channel of communication with them and therefore represented an alternative mode of interaction. In the event under discussion, they took advantage of the playing frame in general, and the Purim periphery of folly associations in particular, and changed the 'rules of the game'. Unlike previous events they did not simply carry out my earlier plan but discussed it openly with me, as an equal member of the group. They asked me to take a role, to become a full participant, and not to remain merely their facilitator, director and spectator as had been my role in earlier events. In this pre-dramatic text we can thus already identify the process of symbolic inversion as a social contradiction through the way in which the elderly inverted themselves from decision-receivers to decision-makers.

This situation can also be looked upon as an inverted 'forum theatre'. The forum is one of Boal's 'theatre of the oppressed' techniques, in which there is first an enactment of a scene presenting an oppression relevant to the particular co-community, then members of this group replace the protagonist (at any point) in the scene and act out an alternative action. Afterwards, there is a discussion.[5] In our case, first there was a discussion and then an enactment. The enactment consisted of two or three different improvisations. Each scene was repeated by different performers according to the participants' suggestions. Below are a few examples quoted from my field notes:

Enactment 1

Performers: Vashti, the queen, played by **Chaya,** sixty-nine years old, plump, friendly, and humorous, one arm and leg are paralysed. The queen's nurse, played by **Eva,** sixty-four years old, a well-preserved, good-looking widow who had lost her husband and two little daughters in the Holocaust and suffers from continuous depression. The queen's hairdresser, played by **Dina,** eighty-four years old, a diabetic widow with a smiling face and a frail voice.

Location: The room of Queen Vashti

Time: Late at night.

The king's messenger has brought Queen Vashti an order from the king to present herself naked in front of his guests. (These are the given circumstances in the biblical story. The actors decided to play the gap between this moment and the queen's answer, which, according to the scroll, was 'no'.)

Vashti: I am not going!

Hairdresser: you don't have a choice (*using her fingers as a comb, taking care of the queen's hair*).

Vashti: But I am a modest woman, it's not dignified to expose my body in front of everyone (*taking the cloth from the table and covering herself*).

Hairdresser: So what are we going to do? Maybe we could send somebody else and disguise her to look like you?

Vashti: That is a good idea! But who will agree to go?

Nurse: I will go, I am not so young anymore, I have nothing to lose...

Vashti: You are so sweet, but if *you* go the king will notice it immediately.

Hairdresser: So what are we going to do? The king will kill us all!

Vashti: I know. Let's wait until the morning. He will sober up, he will be sorry, he will apologize, you'll see...

The following is a replay with a different protagonist. Queen Vashti is played by Dina who was the hairdresser in the above scene. The nurse continues to be played by Eva. Chaya, who was previously the queen, is now the hairdresser.

Vashti: I am not going!

Hairdresser: But he will sentence you to death!

Vashti: I don't care!

Hairdresser: But I care, you are so young.

Nurse: I am willing to die for you.

Vashti: No, no, he knows me well, I am the prettiest woman in the world.

Hairdresser: So what are you going to do?

Vashti: I am not going, no way, he can kill me, I have my pride!

These two improvisations clearly display the two types of symbolic inversion already implied in the pre-dramatic text (as suggested above). The sociological inversion was immediately manifested: Chaya and Dina, two old, handicapped women, played Vashti as very young, very beautiful and very feminine. In both improvisations these features were emphasized not only by the performance of the two actresses, but also by means of the contrast supplied by Eva, who played the nurse as an old lady. In this way the enactment carried a simultaneous twofold message. The explicit, positive one was that given the opportunity the elderly could still enact youth; the second implicit and more pessimistic message was that being old meant being useless and having nothing to live for. This second, implicit message could not be openly stated by the elderly in the centre because the staff would immediately read it as a sign of uncooperativeness and ingratitude towards the immense efforts of the system. The elderly would be labelled as depressed or disengaged, and would be ignored. Inside the protective frame of dramatic playing, however, with the permissiveness of inversion, all messages were equally true.[6]

Therefore, Eva, who played the old nurse, could choose to remain old in the play, and could freely express her personal feeling about old age. Eva played the nurse in exactly the same way both times – as willing to sacrifice herself for the queen and thus demonstrating her generosity through her character. This suggests that old people are not as self-centred as they are often thought to be. They can be very generous and willing to help each other.

In the second version, Dina used the inversion not only to practise youth and femininity, but also to say something about herself in the present. Through symbolic inversion, she got the chance to re-experience some of her old pride in herself. The second kind of inversion, the artistic one, was also already embedded in the first improvisation. In the traditional *purimshpil* Vashti is usually a secondary character and is presented as an ugly, frivolous woman. Here, the elderly actress inverted her to a beautiful, well-educated woman whose drunken husband creates a serious predicament for her. Dina replayed Vashti in a way that emphasized the queen's pride and feminist drive to an even greater extent than had already been demonstrated by Chaya.

Enactment 2
Performers: The king, played by **Menachem**, 75 years old, religious, wheelchair-bound, and small cynical eyes. Counsellor A, played by myself. Counsellor B, played by Ben-David, 75 years old, a newly-widowed, bent, thin man. He has a very wrinkled face, is depressed, and cries easily.
Location: A room in the palace.
Time: Early in the morning.
The king has just sobered up and is conferring with two counsellors. (This situation is not mentioned in the Scroll. The biblical story states that the king was drunk and stupid and sentenced Vashti to death.)
King: I don't know what to do...
Counsellor A: What do you mean you don't know? *You* don't know?! You are God, stick to your words.
King: But I was drunk, I did not know what I was doing.
Counsellor A: So what! You are not going to regret it, what will all the people say, they are just waiting to catch you with your guard down.
King: But I love her!

Counsellor B: You have to forget the whole scene. Vashti was right, she did not want to embarrass you, that is why she did not show up naked.

Counsellor A: She is not O.K. She has secrets with the king's adjutant.

Counsellor B: That's because she is afraid of the punishment she will get.

The King: No! I am not going to kill her, she is a good wife.

Counsellor A: But you can not remain silent! What will all your other girls say? They get above themselves.

Counsellor B: Yes, that is right, but you do not have to kill her – just exile her.

King: That is a better idea, I will think about that.

This enactment was not replayed. The participants agreed that Vashti did not deserve to be killed. They also decided that Menachem was playing the king properly and should not be replaced.

The process of inversion in this enactment was embodied in the elderly participants' decision to establish a role reversal between themselves and me. Although I was a facilitator, they decided to give me a role in the improvisation, but a secondary one – that of Counsellor A. This step enriched the dramatic action with a dominant carnivalesque element – I as a female playing the male in a very exaggerated chauvinistic way. In contrast, the other male characters in the scene were performed by the elderly actors as considerate, thoughtful, caring, loving and understanding of humankind. In the scene they were not afraid to strip me of my usual leadership power. As the counsellor, although I tried in a very pushy way to promote *my* suggestion, my partners chose to take a different strategy. They were tactful and polite, but completely disagreed with me. The elderly actors proved that they still had the social and mental skills to consider different possibilities, to make up their minds and read a given situation properly. If Enactment 1 was a manifestation of femininity, Enactment 2 presented masculinity. This double emphasis on sexuality inverts the accepted image of the elderly as non-sexual beings. The inversive event did not focus on their sexual capability, but it nevertheless created an image of how they conceptualized sexuality at that point in their lives. This portrayal was built up progressively from one improvisation to the other and maintained an unspoken rule: the roles of men and women were never reversed. It seemed in this special event that the conflictual power relationship between men and women, which is so dominant in other stages of life, is of lesser importance in old age. The elderly people's lack of interest in gender role-reversal echoes researches indicating that men and women become androgenous with age, achieving an inner harmony between their 'feminine' and 'masculine' sides.[7]

Menachem did not want to play the king in the traditional way. He felt that the characteristics usually associated with the king, those of stupidity and drunkenness, were grotesque and too close to those imposed upon him by old age. Menachem did not want to reinforce this grotesque image through play but happily took the chance to invert the king into a clever, authoritative man. This creative and courageous move had a strong impact on the participants; they chose not to replace Menachem as the king and he remained in this role for the whole theatrical event. The artistic inversion in Enactment 2 was more dominant than in Enactment 1, and it was also a clear deviation from the traditional *purimshpil*. Menachem changed the biblical story by deciding upon a different fate for Vashti. He demonstrated that a story, even a traditional one, is just a story, and that it can be told in different ways. The relation between the narrative event and the narrated event can be flexible.

Enactment 3

Performers: Mordecai, Esther's uncle, played by **Ben-David** who had previously played Counsellor B. His own first name is also Mordecai. He is holding a newspaper as if it is a holy scroll. Esther is played by a participant whose name is also Esther, her friends call her Estee. She is 78 years old, very fat and plain, and suffers from heart disease. She had the idea of playing Esther as 'a "bimbo", but a really good looking chick'. For the role she decks herself with flowers which she has taken from a vase. Messenger is played by myself, holding a bus ticket in my hands as a scroll.

Location: In front of Mordecai's house.

Time: Morning.

The king's messenger arrives to take Esther. (This situation also does not appear in the biblical story, which tells only that all the beautiful girls

throughout the kingdom were brought to the palace. Esther, the Jewish
girl, is also among them. The elderly chose to play the scene of Esther
being taken to the palace.)
Messenger: Greetings to Mordecai the Jew.
Mordecai: Greetings to the king's messenger!
Messenger: I have an order from King Ahasuerus the great!
Mordecai: We shall hear and obey.
Messenger: The king has heard that there is a beautiful girl in your
house.
Mordecai: Do I have a choice? An order is an order.
Messenger: Good answer. I'll see to it that you get something too.
Call the girl!
Mordecai: Estee!
Esther: Yes unckie.
Mordecai: The king's messenger has come to take you to the
palace.
Esther: Me? How great! What joy!
Messenger: Are you not engaged?
Esther: No! I am young! Take me quick!
Esther is played by another person in the replay. Dina was not happy
with Esther's performance and asked to take the role. Ben-David asked
to replay the role of Mordecai. He was very unhappy with his previous
enactment. 'Now I'll show you!' he said to me, in character.
Messenger: I am the king's messenger, and here is the king's
order: Give me the hand of your beautiful niece, the hand of
Esther!
Mordecai: We have to ask the girl.
Messenger: Why ask? That's a bad habit. Women are proud, we
have to tighten the reins.
Mordecai: I am a Jew and I'll behave according to my custom. We
will call the girl and ask her. Esther come here!
(Dina, who is playing Esther appears humbly before him.)
Messenger: My girl, here stands in front of you the king's
messenger!
Esther: (Sullenly) Yes? O.K.
Messenger: Why such a long face, other girls would dance now.
Esther: I don't have a choice... I will try to be good to the king.
Messenger: Why, aren't you happy? Are you engaged?
Esther: No, but I am happy here!

Messenger: Let us go, we have a long way to go.
Esther: Please let my uncle accompany me, he is so good to me.

In the first enactment Estee exploited the possibilities of inversion fully and happily. She inverted herself to a very young, seductive, frivolous woman. The special occasion freed her from moral constraints, and she could experiment in being a 'bimbo'. In her playing she realized both herself and also Chaya's fantasy. Chaya had confessed at the beginning that, in her role as Esther, she wanted to portray something other than the traditional 'virgin' because she had embodied those qualities all her life. This – in spite of the acceptance of the androgynous nature of old age – nonetheless demonstrates the presence of sexuality in the minds of the elderly, and shows that given the right moment, they may choose to articulate it. Estee enjoyed the freedom of the inversive play and performed her part with gusto. The general tone of the improvisations was what Handelman calls: 'the serious unseriousness of play' (Handelman, 1990, p. 68). Estee, however, brought humour into the event, amusing her friends and thus reminding them of the traditional atmosphere of the occasion. Dina did not approve of this. She performed Esther in the same way that she performed Vashti – as a dignified, modest, well-behaved woman. From our private talks, I also knew that she had been that kind of a woman all her life. She had always been the regulating power behind her husband, but she did this in a modest way. She was proud of her role in 'those good years' and was currently unhappy with herself because she felt she was a 'nobody'. The inverted frame of playing gave her a rare chance to re-experience herself in the way that she liked, and she was unwilling to let the opportunity slip out of her hands.

Mordecai, who played the counsellor once and Mordecai the Jew twice, experienced a developmental process of self-empowerment through his enactments. In his first enactment as a counsellor he had let me do most of the talking and kept himself neutral, trying to make his point to the king without contradicting me explicitly. As he played Mordecai the Jew for the first time, he realized that his manner was too passive. He then used the replay to react more powerfully to authority. In the third enactment, he inverted his character of Mordecai from a coward to a brave man, by using his knowledge of Jewish traditional customs ('I am a Jew and I'll behave

according to my custom. We will call the girl and ask her'). By having the same name as his character, he also inverted himself symbolically from a submissive old man to an activist ('Now I'll show
you!') Ben-David was usually depressed, silent, and introvert. When
he was in a good mood, he used to declare: 'I am a carpenter.' For a
while after the *purimshpil* he began to tell everybody with a smile:
'I am Mordecai, the second to the king.' Then one day he came to
me and protested: 'You have to give me compensation, Purim is
over, you are going to dismiss me, I am no longer the second to
the king.'

Enactment 4

Performers: Esther, played by **Chaya,** who declared at the
beginning of the event that she would play Esther as a 'femme
fatale'. Ahasuerus, played again by Menachem. He holds his walking cane upside down as a sceptre. Referring to his wheelchair,
which he usually hates, he announced in a commanding voice:
'This is made from gold, this is the throne!'
*Esther is brought before the king. (In the biblical text, the king chose
Esther to be his wife. The elderly decide to play this encounter.)*
Esther: (in a sweet voice) Greetings to the king!
Ahasuerus: My maiden, why do you wish to marry me, an old
king?
Esther: The king is not that old, and the king will give me gold,
jewels, and nice dresses.
Ahasuerus: (with fierce and cunning eyes) Do you know what I did
with my former wife?
Esther: What?
Ahasuerus: I had her executed!
Esther: (feigning fear) My king, I shall do anything for you!
I promise to be good! The best!
Ahasuerus: (taking the 'sceptre') With this sceptre I thee wed. Hold
the sceptre, you are blessed.
*(Chaya held the cane 'devotedly' and in the meantime gave a wink and
smile of victory to her fellow participants.)*

In this improvisation, the performers created an encounter between
two familiar images of man and woman – the competent, authoritative, powerful but considerate male and the frivolous, sexy female.

But while Menachem played his part seriously, reproducing his no-longer visible masculinity, Chaya parodied her gender role. She nevertheless showed her audience that, despite the parody, Menachem had taken the bait and was convinced by her feminine wiles.

When the play-acting ended, Menachem, who was generally ill-tempered, gave me a small wink and commanded: 'Take me out!' Usually he wheeled himself out alone. That day, inspired by his enactment as the king, he continued the game informally. I wheeled his chair while he sat up, straight-backed, smiling, and greeting everybody in the corridor. We entered the main dining hall. 'Take me to my place', he commanded again. 'Where are my wives?' he asked, looking around. Chaya, Dina and Estee waved to him. 'Come, sit by me', he asked them. 'You'll have to tell the people about our *purimspil*, otherwise they will not understand what we are doing here with you', said Chaya, laughing. 'Fine with me', said Menachem, and then he looked at me very seriously and said: 'But please not a word to my wife...'

The participants, who had practised symbolic inversion in their own unique mode, found it very difficult to abandon their new inverted roles as king and queen and they continued to hold onto them after the formal end of the event. Empowered by his enactment, Menachem stretched the physical and psychological boundaries of playing beyond its usual frame. Spontaneously using the elastic boundaries of the message 'this is play', he confused his surroundings: was he serious or just pretending? The co-community of the elderly, who often endure a process of deculturalization, demonstrated here that when they do get access to symbolic resources, they know how to make do with it. In this particular case they used the traditional option of cultural inversion not to represent the traditional *purimshpil*, but to create an alternative that articulated their own existential problems in a way that was not open to them in their daily lives.

On the visual-material level, these improvisations introduced ground aesthetics based on formal disproportion – one of the carnivalesque principles (Burke, 1978). This was expressed by the unusual use of the given accessories in the aesthetic space (the given room): the tablecloth, the bus ticket, the flowers, the newspaper, the wheelchair and the walking cane. While the participants felt free to act, bearing in mind the folly-like atmosphere of the festival of Purim, this amusement lay at the periphery of the theatrical event. The improvisations in fact provided a stimulus for a symbolic inversion that created a serious, honest and diverse self-image of old age. The elderly thus managed to create what Handelman notes about public events: 'an occasion which redresses some of the participants' problems' (Handelman, 1990, p. 9), although in this case what helped the elderly to make their point was the rather intimate frame of the theatrical event. Thus, from the social aspect, the participants inverted their low self-images as old people, and their inferior social position in the institution. They created characters that were young, healthy, powerful, wise, functional and humorous. They themselves decided who would play the parts and which episodes would be replayed.

From the literary and dramatic point of view, they did not play the characters from the scroll according to the original text but instead inverted them, changing the basic biblical plot. 'One of the most deeply rooted stereotypes of the aged' notes the anthropologist Haim Hazan, 'is that they are conservative, inflexible, and resistant to change' (Hazan, 1994, p. 28). This *purimshpil* clearly contradicts this stereotype. The elderly took their game of inversion seriously, creating a symbolic protest against their social, cultural and biological situation. The frame of dramatic playing provided them with protection, a means of critique, and the energy to deconstruct tradition according to their own will and enjoyment. The elderly understood the finality and irreversibility of their condition as old and disabled people, but nevertheless demonstrated that they were still capable of articulating themselves authentically. The option of symbolic inversion gave them a chance to display deep feelings and thoughts, a chance that rarely arose in their daily lives in and out of the centre.

Simulating active power had sociological and psychological benefits, as the informal game after the improvisations indicated. The *purimshpil* was a form of a theatrical event that the elderly carried out by themselves and for themselves. It affected them, their self-awareness, and the way in which they perceived each other. The *purimshpil* was a form of dramatic playing that was also a change, and thus the elderly were indeed 'generated as they [were] expressed' (Bruner, 1984, p.9). Through symbolic inversion they could distinguish between their irreversible biological condition and their socio-cultural identity. By using a cultural material as a means to bring up and rework personal materials, they found a symbolic tactic of 'escaping without leaving' (de Certeau, 1984, p. xi), fighting without casualties, and of protesting without risking their rights. They assimilated the subversive power of play to be and not to be at the same time.

3
The Three Elderly Musketeers and their Invention of Play

After a few months facilitating theatre in the day-care centre, I had managed to incorporate almost the entire membership into the theatrical activity. One day, the social worker approached me and, indicating the back left-hand corner of the hall, said, 'Maybe you can try to get the *Iraqi group* "moving" a little.' She was referring to Jacob, Abudy and Sadik, all three of whom had originally immigrated to Israel from Iraq in the 1950s. They used to come to the centre on the same days – Sunday and Wednesday – and sit together at their fixed places, by the same table, in the back left-hand corner of the main dining hall.

They were known for their solitary behaviour and chose not to take part in any recreational or occupational activity. They did not attend the lectures given from time to time on medical or historical subjects, nor did they wish to join my theatre groups. They waited passively for their physiotherapy and for meals. They did not try to communicate with the other inhabitants but were silent most of the time. Even between themselves, Jacob, Abudy and Sadik engaged in very little face-to-face interaction, merely exchanging a few words in Arabic once in a while. From a distance, they brought to mind acting students performing a 'freeze exercise'. I started thinking about how to use dramatic playing as a means of creating a basic give-and-take personal relationship between the three of them. An approach different from the one already in use would be required in order to open up a communicative channel as a basis for a theatrical event. Because the common denominator between dramatic and social interaction is the encounter, which is basically 'a natural unit of social order'

48

(Handelman, 1977, p. xiv), a dramatic encounter needed to be constructed within their own natural space, in the back left-hand corner of the main room. But how to start it and what style to use was not yet clear.

In addition to my formal job as a theatre practitioner, I took on an informal one: helping Shoshana, the woman in charge of cleaning and cooking, to serve the meals – the break, served at 10 a.m. and composed of tea or coffee and a sandwich, and the main meal, served at noon and composed of the clients' orders. These particular events – the meals – appeared to stimulate the greatest reactions on the part of the centre's clients. They were very keen to maintain daily routines regarding where and with whom they would sit, the greetings they exchanged and the food they ate. These characteristics of order, repetition and predictability, which constitute ritual and habitual behaviour in general,[1] partly helped to structure the ceremonial frame of the meals. However, the meals were elevated to the level of a public event by what Mary Douglas (1973) describes as the effectiveness that people assigned to their repetitive activities as a symbolic action to accomplish their desired results.

The elderly clients created an atmosphere of celebration at each meal with the repeated actions unique to those occasions. There was a buzz of happy voices, the exchange of small smiles and friendly conversations about yesterday's meals, the food they were eagerly awaiting and why they chose it, the plentiful food they had at home and the food a relative was going to bring or prepare for them. They enjoyed sharing the personal importance they gave the food; this sharing was a means of presenting themselves to each other as ordinary human beings. This mutual, repetitive play of make-believe invoked in them a feeling of unity, and a sense of belonging.[2] Furthermore, for most of the day while at the centre, these people – chronically sick, in and out of hospital, and socially disengaged from their families and friends – strongly experienced a lack of substantial anchors with which to handle their situation. The act of collective eating had the significance of connecting the meals within the centre with all those other feasts taking place outside, and of linking the older clients back to culture and society, thus validating their existence as normal, accepted human beings.

Eating was one of the activities they could still perform independently, and they devoted a lot of attention to how they carried out the actions, exquisitely prolonging their enjoyment and satisfaction. They were very strict with Shoshana about how their food was to be served. Unwilling to take into consideration the clients' point of view, she found this strictness annoying. The meal rituals created an implicit role reversal: Shoshana, the staff member, who usually demonstrated dominance over the clients, patronizing and even belittling them, was placed in the opposite position at meal times. Shoshana became the 'waitress', the 'servant', while the elderly became the 'customers', the 'masters', with the authority of the last word. They used a repeated fixed text that comprised sentences such as: 'Is this the meal I ordered, Shoshana?'; 'Did you heat my food the way I like it?'; 'I hope you did not forget to remove the skin from my chicken'; 'It is delicious today'; and 'What happened? It so tasteless today!' Transformation, then, was invited at the meal rituals, as Moore and Myerhoff (1977, p. 13) indicate, but could not fully transpire because Shoshana rejected her role. Although a member of the staff, she was at the bottom of the hierarchy. She was uneducated and without professional skills in comparison to the other staff members. She was also a Mizrahi Jewish woman, an immigrant from North Africa, and not like the other staff members, who were oriented towards Western culture. Shoshana was sensitive and tried hard to retain her pride. Therefore, she distanced herself from the suggested role of 'servant' or 'waitress' as she served the meals, responding to the clients laconically, with a poker face. In contrast, I was willing to take up the role of the 'servant' or 'waitress', as I considered that this situation could prove a natural way to get closer to the elderly. By eagerly responding to them, and trying to fulfil their demands, I gained their trust. Later, I could expand the dialogues with them, asking more questions, and receiving new information that might help in planning the next theatrical event.

One Wednesday, when I had finished serving the 10 a.m. break meal, my notes read as follows:

> I took my own tea and sandwich and instead of eating it as usual with the staff, I sat down by the Iraqi trio in the back left corner of the room. 'Good morning', I greeted them and started eating. They nodded at me, suspicious and a little surprised. 'How are you?' I asked. 'All right', answered Abudy. We continued eating quietly [...].

Today [Sunday] I did the same thing. I took my tea and sandwich and sat next to the 'trio'. 'Good morning', I greeted them. They nodded back at me. This time less surprised and suspicious. 'How are you?' I asked them. Abudy answered this time with a smile. 'All right.' We continued eating quietly [...].

This time [Wednesday] when I came to the 'trio' it was after a discussion on the Holocaust, which had taken place in the main hall. 'Good morning', I said to them while sitting down next to them. They nodded back to me smiling. 'How are you?' I asked them. 'All right', answered Abudy, as usual. We ate quietly. 'You know,' said Abudy, 'When the British invaded Iraq in 1941, they let the Arabs attack the Jews. They tortured us and robbed us.' 'Why didn't you say that before, in the discussion?' I asked him. 'Well, I didn't know if I could...' 'Certainly you can!' I said, 'it's important that all of us will know what happened to the eastern Jews too.' And we continued eating quietly [...].

The notes so far reveal how a new, smaller-scale, personal and private ritual became integrated within the larger official 10 a.m. break ritual. This new ritual was built upon a few repetitive activities performed in the same sequence. Some of them – such as serving the food, drinking and eating – were shared with the larger frame of ritual. Abudy, who acted from the beginning as the 'trio' spokesman, elevated himself very naturally to the position of the storyteller. It should be noted that, in this personal and private ritual event, the movement patterns which functioned originally to serve food and for drinking and eating, did not completely lose their original meaning, as Huxley (1914) suggested, nor did they function in 'the background', as Lorenz ([1916] 1976) had indicated. The primary function remained because, for the three participants, as for the rest of the elderly clients, these activities of eating carried a more intensified meaning than that of simply catering to hunger and thirst. As the inventory of activities becomes reduced with age, those activities that are left tend to become very important, not just as instruments, but also as symbols of vitality, the ability to handle situations and the fight against approaching mortality.

Concomitantly, however, another function began to emerge from the repetitive principle: the communicative-sociological one. Both functions, the instrumental, eating and drinking, and the communicative, opening a dialogue, were primary in this particular event during the first stage (described above). But it is from the latter (the communicative) function that play would emerge, as I shall demonstrate shortly. Grimes (1982) distinguishes between ritualization and ritualizing. Ritualization is the process whereby rituals originate. This process is common to human beings and animals and is the process by which non-communicative behaviour patterns evolve into communicative ones.[3] The notes under analysis have so far revealed a process of ritualization. The communicative function is already clear, but the repetitive principle is still functioning without undue notice.

The ritualizing process

If ritualization is the most basic structural pattern of behaviour in both the human and animal realms, ritualizing is a purely human activity: '[T]he process whereby ritual creativity is exercised' (Grimes, 1982, p. 55), and 'The ritual activity becomes self-conscious [...] an attempt to activate, and become aware of, preconscious ritualization processes' (56).

My notes read as follows:

[...] I approached them with my tea and sandwich, they were already following me with their eyes from a distance. 'Good morning', I greeted them with a happy voice. They nodded back at me smiling. 'How are you?' I asked them as usual. 'All right', answered Abudy as always. After a while, he continued to tell me about the Jews in Iraq: 'In those peaceful days (pause) the world was different, people didn't hate each other, we didn't worry (pause). The Arabs loved our prophet's tomb (pause), there was even a Jewish minister of the treasury, Sir Sasson (pause), the bad times started in 1936 when Hitler rose to power. Five hundred Jews were slaughtered within three weeks.' We continued to eat quietly [...].

[On the next occasion] while I was finishing serving the other members, I saw the 'trio' following me with their eyes, waving to me with their hands to hurry up. I took my tea and sandwich and approached them: 'Good morning Musketeers.' They nodded back to me smiling, somewhat amused. 'How are you today?'

I asked them, sipping my tea. 'All right, all right', answered Abudy smiling, drawling his words. We ate quietly. 'You know, we used to go to the night club', said Abudy. 'Watch the belly dancers (pause), only men (pause), you know, there were also some Jewish belly dancers.' Jacob nodded in agreement, with a small, short wink. We continued eating quietly [...].

The first documented encounter reveals the intermediate phase between the ritualization process and the ritualizing process. The repetitive actions are the same but with small variations that show that the participants in the event are beginning to be aware of the ritualization process: the 'trio' followed me with their eyes, and Abudy continues his role as the storyteller more confidently than before. From this stage on, ritual creativity is very clear, as the second documented encounter shows. The 'trio' signal with their hands to speed up the usual process. Abudy carries out his part more consciously, picking up his story exactly where he had stopped the previous time, as though there has been no break in between. Lorenz ([1916]1976) indicates that in the ritualization process some elements of the repetitive patterns are exaggerated, thus giving the action its symbolic significance and producing the playful effect. Abudy, now aware of himself as the spokesman, provides his usual answer, but in a different and humorous manner.

I greet the 'trio' as 'The Musketeers' in a spontaneous speech act. Thinking about the implications later on, back home, I told myself that my mind probably produced this association because they were three men. In fact, however, it was an inverted description of them (as their appearance was far from that of the strong and young Musketeers), although within the new event frame, this description seemed proper – in my eyes, as well as in theirs. In fact, they accepted the fictional role of the three brave knights immediately, without hesitation, which proved that the spark of play had already burst out between us.

The emergence of spontaneous dramatic playing

Within the official 10 a.m. break ritual, another, private, unofficial ritual was processionally constructed. Although the newly engendered activity manifested itself openly, nobody seemed to pay

attention to it. The ritual-within-the-ritual was able to present symbolically what the 'trio' really thought and felt about the centre's
social order. The developing relationship pattern between us symbolized an alternative, more democratic and playful interaction between
clients and staff than was the norm. The implicit seeds of this pattern were already embedded in the larger ritual, as described before,
through the inhabitants' implied invitation for Shoshana's transformation. But transformation needs playing in order to become
manifest: thus, what the large ritual sought was able to emerge only
in the small ritual:

> Sadik and Abudy (Jacob is absent today) greeted me happily,
> with their eyes and hands sending great signals for me to come
> quickly. They seemed eager to start our now rooted ceremony.
> 'Good morning Musketeers', I said to them with a small bow.
> They smiled and indicated gallantly that I sit down. 'How are you
> today', I asked.
> Abudy: All right, fine (pause), Sadik likes to talk to you. You are
> happy, funny (pause), it creates a good mood.
> Sadik: (talking to me for the first time) That's right, that's right.
>   We ate quietly. Sadik started to follow my movements, as if
> he wanted to eat my food, and drink my tea. He tried to sneak
> some of my food. I caught him at it, but instead of scolding
> him I offered him my 'property' as a present. First he took it
> quickly, started to take a bite from the bread, then, as if at the
> last moment, he changed his mind and gave it back to me. We
> continued to eat quietly. Then Abudy said: 'You know, singing was
> not a very respectable profession (pause), there was a Jewish belly
> dancer, she was from a family that had dancing in their blood,
> she was very rich, she used to give money to young Jewish girls,
> so they could get married.'

At this stage, it is becoming evident that as long as the original, basic
ritualistic frame of the small, private ritual remains solid and permanent, the ritualizing process can start to produce pure playing. Sadik
and Abudy elaborated their role as 'Musketeers'; as a response to my
bow, they led me gallantly to my place. As the spokesman, Abudy
interpreted and defined my role as the funny girl and drew Sadik into
the conversation for the first time. Then, he again took up his role as

the storyteller and continued eagerly, with more details, as if the tale had never been interrupted.

Bateson ([1955] 1976) has long argued that play belongs to the sphere of meta-communication. This means that the participants are capable of exchanging signals that carry the message 'this is play', transmitted either verbally or bodily, explicitly or implicitly. This is clearly manifested in the event I present here. Playing was never discussed or formally planned, but because of the high degree of meta-communication between the participants, the transitions between play, ritual and reality were smooth. In this documented encounter the transformative power of play is also clear; it can be seen how play stimulated Sadik to elaborate a spontaneous, non-verbal, raw drama with me.

The process of reduction of the 'possibles' constitutes reality; however, play interrogates this process and opens up 'possibles' (Fink, 1968). The transformative power of play is thus play's capacity to reconnect human beings and the 'possibles' and thus to facilitate the conversion of the 'possible' to the substantial (Atlan, 1994, p. 272). The centre's members, whose average age was 75 and who suffered from physical and mental deterioration, lived in a limited world of 'possibles' in comparison, for example, to young and/or healthy people. Play for the elderly, as the 'trio' demonstrated, could function as an escape route, a way of becoming what they could never be in reality. But, as indicated in the following encounter, the participants could freely produce drama because of the protective and secure frame of ritual they had built up and constantly maintained. My notes read as follows:

> [...] When they saw me coming toward them with my tea and bread they were immediately cheered up. 'Good morning Musketeers', I announced from a distance. They nodded back at me amused, inviting me gallantly to sit down. Jacob: 'Good morning beauty'. 'Ho! With a a compliment like this it's so good to start the week', I answered, flattered in an emphasized way. 'Please, say it again', I asked him in an exaggerated and pleading tone. Jacob repeated his line and we shook hands laughing. Then Sadik started to show signs of jealousy.
> Sadik: ('angrily') What are you doing?!
> I: ('frightened') Nothing, just talking with Jacob.

Sadik: What are you doing with your hands?

I: Nothing, just shaking hands.

Sadik: ('commanding') So, come here!

He pretended to squeeze my hand hard, and I shouted: 'Ai, ai, not so hard.' We repeated this pattern of action three times, as if each time his squeeze was getting firmer. 'You have a pianist's hands', I said to him.

He had very long, white fingers and a light touch.

We started eating. 'So how are you?' I asked them. Abudy answered laughing: 'All right, all right (pause), what do you think of the situation? It's chaos (pause), very difficult in our country now.'

Sadik is showing jealousy again and making faces to me. He again tries to sneak my food from me. I catch him at it. He is becoming afraid of me.

Then I offer him my food. He takes it, examines it carefully, and gives it back. All of us continue to eat quietly.

In this encounter it is evident that the performative actions develop through three autonomous channels. Each member of the 'trio' had his own channel with me. There was almost no inter-action between the three of them. This special system created three different modes of drama. Abudy, as already mentioned, played the storyteller, the spokesman of reason and reality. This role was fixed and he never changed it, only elaborated upon the verbal text. Outside the ritual, he looked shabby, always wearing the same old blue cotton shirt and pants and a beret. He looked like a poor labourer. In the ritual, playing the storyteller and spokes-man, he was transformed immediately into an intelligent, digni-fied person. Jacob always came to the centre wearing a suit, his white hair dampened and brushed carefully backward. He always looked very sad, with his head continuously resting on the table and covered with his hands. In the ritual, he came to life, playing the gentleman-playboy type. It was a short and constant role with no variations, as if constituting the only 'possible' he would ever choose to realize. Sadik, who for a long time was silent and pas-sive, appeared to be a great improviser, especially in pantomime. He always invented new spontaneous raw pieces of theatre, and in time became the real performer of the 'trio'.

At this stage of the developing ritual event, it is also clear that ritual activity functions as a springboard to play, which elaborates itself from time to time into spontaneous drama. The play-drama process then becomes primary, while the ritual becomes secondary. In this particular case, the play-drama process has to do with the content and style of the messages that the 'trio' wanted to communicate, which were not 'indicative' statements but 'subjunctive' ones, always ambiguous, vague and elusive (Turner, 1984, p. 201). These sorts of statements could only be delivered through the medium of play, whose paradoxical message 'this is play' is a meta-message that always declares 'its ongoing self-contradictions and self-negations' (Handelman, 1990, p. 69). Thus, the 'trio' could display a serious unseriousness which was not only the suitable and safe mode of articulation for them, but also a metaphor for their social and existential state in the lived-in world, as the next incident demonstrates:

> [...] we all laughed, then Sadik winked at me and said: 'ask Sara why she isn't giving me some work'. Sara, the occupational therapist, was striding past us. I asked her in a very low and deep voice: 'Sara, why aren't you giving some work to Sadik!'
> Sara: (joining in the play) Because he doesn't deserve any...
> Sadik: (crying like a baby).
> Sara: O.K., I'll forgive you, here, I'll give you some work.
> Sadik: How much will you pay me?
> Sara: The same as I pay the others (everybody works for free).
> Sadik: I don't want to be like everybody else.
> Sara: So I'll give you two coins a day.
> Sadik: (gestures as if he is going to beat her up).

The play between Sadik and Sara enabled Sadik to express his thoughts and feelings about the normative power structure within the centre. The staff was composed of domineering, European, 'superior' women (except Shoshana, the cook and cleaner). Sara was about fifty years old, fit, well-dressed, wearing quite expensive jewellery. She had at one time worked in a kindergarten and thought that treating old people as children was the proper way. Like all the other authoritative women, she thought she was doing a marvellous job, and that all the centre's clients should be grateful. Most of the elderly men spent a lot of their time in a small, closed room where they worked at fitting together series of tiny screws. They did not receive any payment but the female authorities nonetheless expected them to feel productive. Sadik, an emigrant from an Arab country who had spent his entire life in a male-oriented society, found this role reversal in his later life to be very uncomfortable. It was our private ritual that energized him with the stimulus of playing, so that he could produce a piece of raw theatre and use it as a means of self-empowerment to protest against the centre's official social order. The paradoxical state of the play medium created a special mode of struggle, which was suitable for Sadik. He could articulate his feelings without paying the heavy cost of being penalized by exclusion from the centre's benefits.

The transformation of drama back to ritual

When I entered the centre, I saw Sadik, supported by the nurse, walking slowly to the toilets. When he noticed me he started to make funny faces at me. Then he put on a very cross face and said to me while waving with his stick: 'You see this stick, I brought it for you, I'll beat you up!' He continued to keep his angry face for a short while and then burst into laughter. Later in the morning, after the '10 o'clock break' ritual had started, he waved me over to him. When I approached him and his friends he turned his face as if he had not called me and did not even know me. Then he turned his face back to me and the minute our eyes met, he again turned his face sharply from me. After he had done this a few times, he started laughing, holding out his right hand to shake mine. We played the 'squeeze game': he squeezed my hand 'strongly', and I reacted 'Ai, it's hard.' We repeated this pattern twice more, each time with him squeezing my hand more 'firmly' and I reacting more 'painfully'. Then we all laughed. After that I went back to my job and served the meal. The 'trio' watched me the whole time, making signs to me to hurry up. 'Hello Musketeers', I said, coming up to them with my tea and sand- wich. 'Hello beauty', said Jacob. 'Ho, with such a compliment it's so nice to start the week. Please, say it again', I said. 'Good morning, beauty', Jacob repeated. Sadik put on his jealous face. We ate quietly. 'So, how are you today?' I asked them. 'All right, all right', replied Abudy in his usual manner. 'So, what's going on?' I asked again after a pause.

Sadik: The world is swinging (he is demonstrating with his hands).

I: Until when?

Sadik: Until the Messiah will come...

Abudy: The Messiah (pause), yes, believe me, the Messiah could have been here if people were not so bad (pause), thieves...

Sadik: What nice shoes you have, buy me some too (he is pointing at my pink plastic sandals).

I: Very well, I'll buy you exactly the same.

Sadik: When? (pause), when the Messiah will come...

We all laughed, then we continued to eat quietly. Suddenly, Sadik sneaked my food and we played the 'thief game'. Then we continued to eat quietly.

When I came in today I noticed that Abudy was absent. The minute Sadik saw me, although I stood at a distance, he started playing with me the 'I don't see you' game. Then he made a few funny faces and smiled, as if releasing me to my formal work. When I approached Sadik and Jacob later, with my tea and sandwich, the usual ritual took place. I greeted them as Musketeers. They nodded back at me smiling. Jacob called me a beauty, twice, as usual, and Sadik became very 'jealous' and then played the 'squeezing game'.

While we ate quietly Jacob leaned his head on the table.

I: Here is Joseph the dreamer.

Sadik: (playing Joseph) I had a dream (pause), I had a dream...

Jacob: (raising his head and speaking in a biblical tone): I had a dream and here is Stella...

I: And who is Stella?

Jacob: (quietly) She is a beauty, married to an old doctor, and here is Sasson, the young dancer, she is inviting him to come to her (pause).

Once I had a good life (pause), women ...

In the meantime Sadik was humming, making signs to me to give him my attention. I showed them my new shoes.

Jacob: Nice, they suit you.

Sadik: *Alte Zachen* [old clothes, an expression in Yiddish].

I: (in an insulted tone) Not true.

Sadik: Sure, here, see, Jacob is already dreaming about them. (Jacob leaned his head on the table).

Sadik: (in a biblical tone) The dreamer...

Jacob: (continues the phrase) And here is a 'ladder with its head in the sky...'

Sadik: So, when will the Messiah come?

I: Maybe Jacob will dream about it.

Sadik: No, no, he is dreaming only about beautiful girls...

Then we ate quietly. Before I had finished my food Sadik played the 'thief game' again and as a farewell we played the 'squeezing game' again. This time he added: 'Why are you screaming, it doesn't hurt this time', as if the other times he had indeed hurt me...

When the ritual frame is firmly established, play can spread even beyond it. The ritualistic activity creates the safety zone, the sheltered base that first encouraged the 'trio' to play with me and to test their symbolic attack on me. In time, seeing the success of his actions, Sadik took the opportunity to organize 'guerrilla attacks' outside the ritual frame, using play as a cover. This tactic is manifested in his play with Sara and also in the above encounter. He elaborates upon a new game the moment he sees me, transforming his stick into a sign of power and through this action restoring the role reversal found at the centre to its 'proper' state. Now it is he who has the authority over me; it is he who treats me like a little child. Later, being still empowered by the ongoing private ritual event, he does not wait for its regular moment to start. He insists on playing in advance his new 'I don't see you' game. Here, he chooses a children's game, using a childish omnipotence to make me, and not himself, invisible. By playfully and exaggeratedly ignoring me, he causes me to feel neglected, as if I was not there. I think that this feeling of invisibility, which so often characterizes old age, was probably one of Sadik's strong feelings of insult. Although he was incapable of moving by himself, he never asked for help but instead waited for the staff to attend to him. Usually, of course, the staff did not respond quickly, even if a client called them. Immediately after this new game, Sadik repeats a familiar one and thus makes do with the ritual structure to incorporate the new into the old, to stretch the protective cover over the actions that otherwise might seem too abrasive in the reception process.

The old, basic ritual pattern is thus always present, but what makes it fresh and gives it new content is the process by which the drama from yesterday's ritual becomes ritualistic in today's ritual. Through the repetitive principle and its own transformative power, play re-forms drama back into ritual, and enriches the ritual event. The instances of ritual, play and drama take place simultaneously but in harmony. Play and reality alternate, dissolve into one another and separate. Through this cyclical, metaphorical manipulation of time, the participants temporarily overcome the constraints of chronological time and create anew their personal time. The 'trio' appeared at first to have given up the fight. Nevertheless, through the ritual-drama-ritual frame they showed that their life resources were still intact. They needed only the right stimulus to use them creatively and to bring significance to their daily routine.

The repetitive principle which organizes ritual is usually associated with conventionality and rigidity. But, as the three elderly Musketeers revealed, the repetitive principle holds within itself the potential for creativity and playfulness. When the boundaries of one frame of symbolic activity are steady and clear, there is the possibility of creating an additional and different frame of activity. The ritual frame, in this specific case, was the trigger to play. Functioning as a safety barrier, ritual enabled the emergence of spontaneous, informal drama. With time, this playful behaviour, determined by the same repetitive principle, turned back into ritual behaviour. Play, the common component of energy between ritual and drama, was the medium that enabled this process of transformation, keeping the event always flowing and the participants always changing. Whereas the centre's daily routine was not able to supply the three older clients with enough opportunities for transformation, the ritual drama event became a communicative resource, a meta-language through which they could articulate themselves while subverting and temporarily overturning the centre's formal, official routine. They communicated messages that could not be transmitted easily outside the frame of the event. The 'trio' needed the centre's services; because they were chronically ill (Sadik and Abudy suffered from heart disease, and Jacob was recovering from a stroke), their means either to alter their situation or openly to oppose the social system were very limited. Empowered by play, they were able to create hypothetical situations. Thus, their symbolic, performative activity constituted a change, effecting as well as affecting the quality of their life.

The ritual drama event was an emergent, elusive activity, which took form through the process of its creation, a 'proto-event' (Handelman, 1990, p. 20), never named by the participants and never reaching a fixed form and content. The basic pattern of ritual-play-drama-ritual acted as an open-ended creative process that not only reorganized linear time into cyclical time but also cancelled time's significance by constructing a different, multiple time. Each ritual event revived the previous one, continuing it without any apparent break. In every event, the participants moved smoothly from past to future or present, sometimes even being simultaneously in more than one time, when they carried out their different actions all together. Through their unique form of playing the three elderly performers defied two common premises: that about the old sticking to ritualism and thereby not liking and/or not wanting and/or being unable to play; and that about ritual and drama, perceived as differentiated activities in modern society. The former is conventional while the latter is creative.

Old people are indeed attached to rituals and arrange their daily life around habitual activities; but then all people need rituals in order to attain a sense of unity, continuity, order, security and dominance. Young and healthy people, however, are engaged in so many activities that their ritualistic activity is less apparent than with the old and the sick. The latter – especially those in old-age homes and rehabilitation centres – being deprived of their social roles and status, give the impression that their lives are purely ritualistic.[4] In the post-industrial, hi-tech world, where things tend to change even faster than human perception, old people may find it very difficult to catch up. When this happens and they feel socially disengaged, accustomed habits and activities can compensate to the extent that routine is elevated to the sphere of ritual, as was demonstrated by the elderly clients at the centre.

Ritualistic behaviour was performed by the three elderly men as a tactic to symbolically overcome their anxiety. 'Paradoxically', as Barbara Myerhoff notes, 'it [ritual] uses repetition to deny the empty repetitiveness of unremarked, unattended human and social experience'

(Myerhoff, 1984, p.173). But the three elderly players pushed and challenged this discourse beyond its well-established boundaries, showing very clearly that old people as ritual practitioners are also potentially good candidates for play. Where there is a strong orientation toward ritual, spontaneous drama may emerge. Sadik, Abudy and Jacob's circuit from ritual to drama and back also proved in practice what theories of play and ritual have longed formulated: ritual and play are not necessarily polar activities but in fact share basic characteristics.[5] Moreover, these elderly men's spontaneous drama produced raw theatrical events that manifested the transformative power of play, keeping the axis of options for them ever open. Thus, the elderly performers made do with ritual and became the 'trio' – the 'three Musketeers'.

4
Playing the World-Upside-Down at a Children's Medical Centre

Childhood, like old age, is an oppressive socio-cultural construction. Although the modern invented category of childhood has gained increasing visibility and attention, children, like the old, are nonetheless stigmatized as weak, dependent and selfish. It is fortunate that while the elderly comprise an essential and permanent co-culture that signifies death, children are the signifiers of life and only temporarily positioned as a co-culture (Hazan, 2006a). Hospitalization, however, forces children to cope with a strange and painful environment that transforms them into a traumatic co-community. This chapter focuses on how hospitalized children together with theatre students made do with 'carnivalesque enactment' in a children's medical centre. Before discussing the theatrical event itself I shall first elaborate upon carnivalesque enactment as a unique form of acting.

Carnivalesque enactment as a type of acting

Approaches to acting usually fall between the extremes of naturalism and alienation: the affective and cathartic or the cognitive and didactic. The former approach argues that acting must offer a sense of authenticity, while the latter emphasizes the artificiality of acting. Diana Delvin breaks this binary perception by presenting a third, intermediate mode of acting entitled 'acting as play'. Here the emphasis is put not on the actor's skills but on the actor's presence, and on the knowledge shared by audience and actors that the play presented is only a game (Delvin, 1989, p. 103). I would add that the playing approach to acting can be found in those performances in which the performer is *playing with acting rather than acting properly*. Carnivalesque enactment is a particular type of such playing that has its source in the heyday of carnival in medieval and Renaissance times. There are still carnivals all over the modern world, among them the famous carnival of Rio de Janeiro, the Fasnacht carnival of Basel in Switzerland, the Ness carnival in France, the Mardi Gras carnival of New Orleans and the Venice Carnival. Each has its own unique features and history, indicating that carnival has always held multiple forms that vary according to both past and present context. Moreover, the term carnival is currently often used to describe any festival, celebration or party that contains masquerade. In order to outline some unique qualities of carnival as a socio-cultural phenomenon, it is therefore necessary to go back to its roots.

In medieval and Renaissance times, when society was built upon a general, unified philosophy of life and a shared ceremonial cultural consciousness, all carnivals expressed themselves through the main principle of a World-Upside-Down. This fundamental conception was based on a dual vision of the world and society, a double condition in which one facet is the daily reality, the world of order and seriousness, and the other is an inverted world of laughter and folly. Carnival was thus a licensed inversion of bipolar oppositions (Bakhtin, 1968, p. 6), which was expected by the authorities to express social solidarity and harmless laughter, but was in practice a form of subversion. The performative manifestation of the World-Upside-Down principle was through playing in general and role reversal in particular. Carnival, it is important to note, was a liminal phenomenon both in terms of its playing display, and in its position in time and space as a festive event rupturing mundane reality. Moreover, its particularity as a special mode of playing is revealed through its liminal, material appearance, which distinguishes role reversal from other forms of playing.

Carnival is immediately associated with disguise, but carnival's dressing up is different from that in play, ritual and theatre. Dressing up in carnival articulated role reversal as a clear sign that a departure from reality had begun and that the social world was being temporarily inverted. 'Carnival', as Bakhtin observes, 'celebrates change itself, the very process of replaceability, rather than that which is replaced' (Bakhtin, 1973, p. 103). Continuing this idea, I would say that carnival manifested the *process* of inversion rather than the inversion itself, that is, not an inverted reality in its complete form, but the inversion in its liminal stage, in the process of its becoming. Dressing up in carnival thus signalled its liminal, chaotic unformulated, disordered phase and realized it through three characteristics: (a) formalistic disproportion, represented for example by gigantic noses, horns and enormous phalluses; (b) powerful, strange and colourful disharmony; and (c) a mixing of incongruous and incompatible elements, such as a man dressed in woman's clothing and wearing a cow's tail in one of the English medieval carnivals (Chambers, [1903]1963, p. 210). The implausible combination of elements of the costume and mask were the performative means by which cultural components were re-decoded simultaneously to transmit both polysemic meaning and non-meaning, thus putting the normative cultural codes into a reflexive and sceptical position. Within the overall ideology of turning the world upside-down, unsuitability, dissimilarity and incongruity became logical and meaningful. My contention is that this exceptional design system calls for a similar, suitable kind of enactment. Just as there was no congruence within the performer's costume itself, there was also no full correspondence between the performer's appearance and conduct. This kind of performing reflected a total departure not only from normal everyday behaviour, but also from what was considered to be normative theatre enactment, that is, an enactment in which compatibility between the character's external appearance and his activities helped in the creation of a coherent meaning.

Carnival offered a different, deliberately muddled system of costume and behaviour that appeared disoriented and preposterous in relation to normative behaviour, both in daily life and theatre, but that was nonetheless intrinsic to the world of carnival itself. This philosophy celebrated the temporary process of turning the world upside-down, but did not constitute any serious attempt to construct a permanent alternative for the social order. Therefore, the particularity of carnivalesque enactment lay in its liminal mode of role reversal, which generated an incoherent sequence of nonsense activities played by undefinable characters, manifesting liminality through the absence of coordination between role and costume and through inconsistencies within the role itself.[1]

Carnivalesque enactment, to conclude this theoretical exposition, is a type of performing based on incoherent and absurd dramatic playing. The performer's unsettling mixture of activities, like his eclectic costume, present a process of fictional character creation that continually subverts expectations. This type of acting is enormous fun for the performer who is enabled by it to enjoy the process of deconstructing human behaviour, both as evidenced in daily life and in the theatre.

Carnivalesque enactment and therapy

The socio-therapeutic intention of carnival was to facilitate a licensed safety valve for releasing social agitation and unrest (Burke, 1978). Appropriated by young patients, this safety valve may operate on two levels. It can stimulate both an emotional discharge and a cognitive educative process. Carnivalesque enactment is a tactic through which the young patient may reach a deeper awareness and insight of hospital reality as well as a better understanding of its significance to his or her own life. By exercising such subversive devices as role reversal, formalistic disproportion and colour disharmony, the child has the chance to deconstruct the components of the hospital universe in a variety of ways, learning both to disarm them as threats and to appreciate their urgency for his or her well-being. As both spectator and participant in the carnivalesque enactment the child is enabled to invert serious situations and to act out inevitable anxieties. The theatrical event brings with it a temporary feeling of power and control that can equip the child with the mental resources to handle unpleasant treatment.

Choosing the participative model of performance at the hospital

In 1995 a theatre project was carried out at the Rabin Children's Medical Centre in Israel by students from the Theatre Department of Tel Aviv University. The Rabin Children's Medical Centre was founded in 1992 and welcomes sick children from Israel and the Middle East, with no restrictions relating to religion, race or nationality. The hospital's philosophy is that the needs of a child in hospital differ from those of an adult. Children become ill more frequently than adults;

they need special medical appliances that suit their small size; and they react acutely to hospitalization, often being anxious and fearful of being abandoned by their families in a threatening environment (Katzanelson, 1993). Hospitalized children are therefore a special, temporary co-community and in order to alleviate their symptoms the Rabin hospital employs a variety of strategies. Architecturally, it is a huge light-filled building, with translucent elevators, candy stores on the entrance floor and multi-coloured mobiles hanging from the ceiling. Pink, purple and green are the dominant colours in the rooms, the circular corridors, the staff uniforms and the children's pyjamas. Treatment is carried out exclusively in a specially equipped room, and the area containing the children's beds is made as homely as possible. There are adjacent beds for the parents, who may remain with their children day and night. The menu is varied, offering the children a choice of chips, pizza, hamburger and other fun foods. Each ward has a big playroom, computers and a library. Through this therapeutic approach the staff tries to narrow the gap between the outside world and the hospital, and to create a protective and pleasant environment. Out of this policy had arisen the idea to invite staff and students from the Theatre Department of Tel Aviv University to work with some of the children.

We began with two facilitators and four students, who were later divided into two teams, each in a different ward. I worked as a facilitator with two students in the internal medicine department for eleven months. As a former actress and a community-based theatre practitioner, I was able to guide the students to accomplish their dramatic ideas. According to the basic model of community-based theatre in Israel, the facilitator works with the same group of participants over a long period of time. At the beginning of the activity, the facilitator aims to establish an atmosphere of friendship and trust within the group through different dramatic exercises. The relevant personal materials raised during the activity are then used as the basic resource for a play written by the group, assisted by the facilitator. This play, staged by the facilitator and performed by the participants, is then presented in front of the local audience.

In searching for an alternative model for the hospital, the students and I decided on a 'participatory (participative)-interactive performance'. They would prepare a written text in advance, which would deal with a specific topic related to the hospital environment. This topic would present a clear conflict with the manner of its denouement, with the children invited to play games and to take active roles at certain, predefined points of the event. I advised the students first to examine their associations in relation to the hospital, which they perceived as a huge mall, a kindergarten or an amusement park. I guided them to follow these associations while creating a theatrical event for the next visit.

Realistic enactment combining participation through games

The first theatrical event was called *The Flying Bed*. It started with an informal prologue, in which the students in character entered each room singing a song and inviting the children to the playroom. *The Flying Bed* began as an improvisation, but within a month had consolidated into a short play in which four characters arrive at a strange place and try to confront their new situation. Sulky, the boy who is never satisfied, wears overalls and a cap and his face is smeared with make-up; Beauty, the girl who is always busy with her looks, wears flowered pants and a shirt, has a ribbon in her hair, and holds a brush and a mirror; Logy, the over-rational kid, wears a short overall, long socks and spectacles and holds an abacus; and Tricky the trickster wears a track suit and holds various tricks. The bed is a hospital bed covered with colourful ornaments, bells and antennae. At the core of the performance are two particular scenes: Beauty's efforts as an outsider to become a part of the flying group, and the arrival of the band in a 'strange land', the hospital.

The students shaped the four characters naturalistically and acted them out using a realistic style of acting. This process happened almost automatically, as the Stanislavski method was the dominant acting style they had acquired at the university. The children's participation was planned for the second episode and based on the game structure. The idea was to stimulate participation through simple tasks and activities, rather than through acting out dramatic characters. The student performers asked the children various questions about the hospital and then played games with them. The first game was the 'animal game', in which one performer called out the names of animals, plants, colours and so on and the children were supposed to clap their hands only when hearing the name of an animal.

The second game was 'puffing puff', in which one performer stood with his eyes closed while the children crept silently towards him. When another performer called out 'puffing puff!' the first performer would open his eyes while the children 'froze'. To close the performance the performers assisted the kids in making different animal masks.

The children received the performance indifferently, with a low level of attention, and their eyes wandered around the playroom. In the second, participatory, phase they began to show more interest, and some of them even became willingly and voluntarily active. Analysing this event afterwards, both the students and I had the feeling that the performance had not worked. The passage from the dramatic text to the games was too sharp, cutting the performance into two unrelated and disconnected parts and confusing the children in their reception process of the whole event. Moreover, while the students had intended to cheer up the children by drawing them into a distanced land of fantasy and games the overall tone of the event had nonetheless been quite serious and thus too like the given reality.

The second theatrical event, *Longing*, enacted the story of a little girl named Effi. She is hospitalized and feels homesick. She tells her pillow and the children in the audience about her beloved toys at home. When she falls asleep the toys come to visit her and play with her and the other children. I advised the students not to postpone the children's engagement until the end of the show, but to spread it over the entire event, and also to try to introduce an element of 'drama' into the games. Several activities were planned for the young patients. Upon entering the playroom each child received a sticker. When Effie fell asleep and the pillow (the other actress) woke up, she asked the children to come and decorate her with the stickers to become a magic pillow. Then, in order to start the magic dream, the children had to sing together 'Ding, Dong', simulating the sound of many small bells. In 'dream land' with Effie, the magic pillow then played the 'Make them laugh' game with the children. She made funny faces and any child who laughed had to join her and assist her with more funny faces. Then, before transforming herself into one of Effie's beloved toys, she asked the children to show in pantomime the toy that they missed the most. At the end of the event the children were invited to draw their beloved toys. The children were very enthusiastic about receiving the stickers and waited eagerly for the cue to get up and stick them all over the pillow. It became clear that the part that the children found most amusing was when they could get closer, and even touch the performer in role as if she were a pillow. They became highly absorbed in pantomiming their beloved toy, and almost all of them agreed to stand in the performing space, in front of the others and their parents, and demonstrate their own favourite toy.

In the subsequent analysis of the *Longing* encounters, the students noted that the children's involvement in this theatrical event had been more extensive and eager. However, there was still something missing, resulting in an over-didactic and over-naturalistic perform-ance. I suggested that they reconsider their acting style and look for an alternative.

Carnivalesque enactment in motion

I advised the students to reconsider their first associations with the hospital and try to focus on the most representative. The students again noted that it did not look like a hospital but more like an amusement park, or a circus. 'But it is still a hospital', I insisted. 'A nurse is a nurse. A doctor but a doctor, and for the child it is still a frightening place.' This comment stimulated them to notice the discrepancy between the hospital's physical appearance and its real activities. I asked the students to think in images. 'The hospital has dressed up', they said, 'but under its circus disguise it still functions as a hospital.' This image inspired them to realize their role as drama agents in the hospital. They should energize this 'frozen circus', bring it to life, and through their enactment complete the image that the hospital ideology wanted to create. The unique therapeutic policy of this hospital indeed needed a special dramatic enactment in order to evoke its new, positive medical image in the children's minds. This new premise was the point of breakthrough, and during the next two performances a new style of enactment was developed.

The third theatrical event was called *Acamoly and Infusy* (aspirin and infusion). Acamoly wore a big round sign on her chest and back inscribed, 'Acamoly'. Beneath she wore huge pants with a colourful big shirt, slippers, ribbons, a plastic squeaky hammer and a whistle. She had white make-up around her eyes and a big red and white mouth. Infusy wore a green gown, and held an infusion stand decorated with ribbons and balloons. She had ribbons in her hair, a whistle on her chest and a bottle of water in her other hand. Beneath the green gown she wore a doctor's robe with a stethoscope and huge spectacles hung over her chest. Such costumes are very similar to those of the carnival described earlier, in that they too have formalistic disproportion, colourful disharmony, and a mixing of incongruous and unexpected features. The appearances of these characters stimulated a lot of laughter and amusement. It immediately created a sense of ease in the children, and a strong urge in them to approach and examine these strange individuals. At the first meeting between Acamoly and Infusy, Infusy sang discordantly, while Acamoly complained about her headache. Infusy offered her an aspirin. Acamoly, feeling insulted, provoked Infusy and asked her what she had today in her infusion bag: Coca-Cola or Fanta. Infusy hit back and answered that she was really thirsty and opened the bottle of Fanta. While she was bringing the bottle to her mouth, Acamoly asked her, as if very seriously, what time it was. Infusy turned her hand with the bottle to look at her watch and the drink spilled out. This gag was repeated twice. Then Acamoly said she had a secret she could not tell. Infusy began to sing croakily again and in order to stop this awful noise Acamoly had no other choice but to tell the secret...

In this scene carnivalesque enactment has already begun, manifesting the way in which carnival is a unique form of playful acting. The two performers simultaneously enacted two clowns, two gossiping neighbours and two medical products. In the second scene, the playful acting became intensified when role reversal, another prominent carnivalesque principle, became eminently clear. Acamoly told Infusy that Ronith, a young patient, had not let Doctor Danna treat her, and instead had asked her to exchange roles! Now, in the performance, Acamoly took the role of Ronith and Infusy took the role of Doctor Danna. Ronith (Acamoly) persuaded Doctor Danna (Infusy) to play a game with her. In the game Ronith acted out the role of the Doctor and the Doctor played a little girl. Ronith examined the doctor very seriously with her plastic hammer, tapping her all over her body, pulling long coloured strings from her mouth, and asking her to repeat strange sounds. At this stage Ronith/Doctor invited the children to come and play the Doctor. In this scene the social significance of role reversal could be traced. Every young player got a green gown, huge spectacles, an enormous hypodermic, plasters, bandages, a plastic hammer and all kinds of funny clownish accessories. Each child in turn checked the patient very seriously, tapping her all over her body, pulling reams of coloured string from her mouth, giving her injections, plastering her mouth and bandaging her legs together. The Doctor as a patient reacted hysterically, begging to go back to her former role. The children tickled her and asked her to repeat strange sounds after them.

The children greatly enjoyed their participation and wanted to prolong it for as long as possible. Each child had to give a remedy, which was generally to eat candies and play in the playroom. Doctor Dana, quite overwhelmed by the 'treatment', got her part back and told Ronith that she hoped she and the children had now overcome their fear of her. This carnivalesque atmosphere encouraged the children willingly and enthusiastically to participate in the drama. The student performers no longer played realistically, but in an exaggerated, caricatured and disordered manner, which drew the children spontaneously to imitate them and thus to become actors like them. The children enjoyed the idea that they could give orders not only to the performers but also to the 'Doctor'. The opportunity, even if only temporarily, to turn the hospital world upside-down empowered the children with a feeling of control over the frightening medical characters and treatments. For a limited period of time, patients, actors, doctors, nurses, clownish accessories and medical equipment came together, as if belonging to the same entertaining and funny family. Carnivalesque enactment was responsible for the humour and laughter that had been missing from the previous performance, but it also enabled the children to look at the medical components of their present lives from different points of view, and to perceive, if only partially, the importance that they held for their lives. After this performance they were more cooperative in writing down their wishes concerning their medical treatment. One child wrote that he wished the treatments could be combined with plays, like in the show. Another child wrote that he wished the doctor would give him a warning sign whenever it was about to hurt. Yet another asked the doctor not to lie to him, telling that he was only going to play and tickle him and that would be the whole treatment. The painful treatments became less demonic, being perceived from a more rational angle and with different insight and perspective. A particularly practical outcome was that the children's requests were hung on the board in the treatment room so that the nurses and doctors could read them.

Carnivalesque enactment developed further in the fourth theatrical event, called *Mister Fun*, in which two detectives, Mr Det and Mr Ective help the children to find Mr Fun in the hospital. The two performers played simultaneously the parts of the detectives, a crying girl, a nurse with an enormous hypodermic in her hand, an absent-minded doctor, Acamoly, Infusy and Mr Fun. The performance was manifested in the form of a mock inspection. The detectives guided the children through a quick course in detecting and then invited them to inspect the nervous nurse with her huge hypodermic, the absent-minded doctor, Acamoly and Infusy – the 'terrified suspects'. The children played the leading roles, while the performers played the 'terrified suspects'. The children in role determined the course of the performance and were responsible for the entire event. In each encounter they quickly understood that as long as they continued to become more and more active, amused and happy, they themselves caused the theatrical event to give birth to Mr Fun. At the high point of the event, one of the performers transformed himself into a funny, happy, jumpy gentleman who brought the children new ideas for having fun. The premise was that the potential for fun lies within the children, but that they need the theatrical show to stimulate and produce it. After this performance the children started to greet the performers as 'Mr Fun' or 'my friend'. One mother told us that her daughter would not leave her bed except to see 'Mr Fun'. Other mothers joined the students and children parading after the performance into the corridors and rooms, to tell those children who could not attend the performance all about 'Mr Fun'. The teacher who came in to teach the young patients twice a week, in the morning, came one afternoon specially to meet 'Mr Fun', having heard the children speak enthusiastically about him. The social worker, who acted as the formal representative of the hospital, and with whom we were in constant contact, decided to invite all the medical staff to see the performance and gain inspiration. Following 'Mr Fun' she proposed the idea of involving the funny characters in the instruction sessions held with children before heart surgery.

Carnivalesque enactment as a tactic for articulation

In this chapter, I have suggested the efficacy of carnivalesque enactment for the empowering of hospitalized children. The therapeutic potential of humour and play is already a well-known and widely documented issue. It has led several hospitals, such as the Good Samaritan hospital in Los Angeles, to introduce a special humour channel into the ward televisions. The Cancer Treatment Center in North Carolina offers a cart full of humorous materials. Instead of flowers or chocolate, one can give a sick child a book of caricatures or jokes, a tape with funny skits or a comic videocassette. At the Sheba Hospital in Israel, young patients are prepared for surgery in a playroom. A trained nurse role-plays with the children using various surgical appliances. By playing with real medical equipment, it is hoped that the young patient will approach the surgical event more calmly and with less potential for trauma. At the Rabin Medical Centre, nurses use Playmobile building bricks in order to make the surgery room for the sick children both a real and a less threatening place (Katzanelson, 1993).[2]

As we have seen, the unique therapeutic potential of carnivalesque enactment in a children's hospital goes beyond its ability to facilitate humour, laughter and play. Children entering the semi-closed world of the hospital, with its restricted and unfamiliar rules, are immediately brought into a temporary co-community that suffers from inarticulation. The adoption of carnivalesque enactment can reduce the threatening and alien effect of unfamiliar medical instruments. It also humanizes the authoritative, all-powerful doctors and nurses by temporarily making them appear small and helpless, while the children, in contrast, give the orders and perform the various strange treatments. At the same time, the special style of costume and enactment continually reinforces the children's understanding that this is indeed only a temporary reversal that does not and should not cancel the treatment itself. The therapeutic effect of such enactment derives from the existence of a protective frame whose incoherence and absurdity clearly indicates that for a limited time – inside the frame, and only inside it – the children, guided by the performers, make do with theatre and turn the world of the hospital upside-down. Carnivalesque enactment has the potential to make the strange familiar and to mitigate frightening hospital stereotypes.

It provides the young patients with the opportunity to transform themselves and their surroundings into a more endurable reality. Within such an inverted world, children can articulate their visibility and presence as active subjects who control the space which they occupy. Feeling empowered, they may then accept the painful treatments more easily and reasonably.

I would like to underline here the value of carnivalesque enactment, beyond the frame of the hospital, as a tactic for articulation that has developed 'from below' by 'the people' and for 'the people'. It enables the co-community to invert different hegemonic hierarchies into mixed and hybrid symbolic worlds, and to imagine the social reality as multiple, over- or under-sized, disproportionate, off-balance and incomplete. The Janus-faced carnivalesque enactment as both a licensed socio-political disruption of the hegemony and a potential practice of resistance suits the co-community, which always suffers from a complex relationship with the establishment. Theatre projects do not usually emerge organically from below but are instigated from above by various local or state bodies for the interpellation[3] of the co-community. Carnivalesque enactment on the other hand enables the co-community to articulate criticism and protest without taking dangerous risks. While retextualizing the social reality and exposing its discrepancies, the co-community also reaffirms its desire to become an active agent of this same social order.

5
'Theatre of the People': Rhetoric versus an Apparatus for Subversion and Control in the Mizrahi Co-Community

Theatre in ethnic co-communities began in Israel in the early 1970s, mainly within the Mizrahi co-culture (Jews originating from Arab/Muslim countries) and has since become known as 'community-based theatre'. This form of theatre is generally initiated and sponsored by municipal and/or state welfare and cultural bodies as a means for interpellation. The institutional agencies assume that the participants' self-expression through theatre will facilitate or testify to their integration into the dominant order. In contrast, from the bottom-up perspective of the theatre group, the theatre project provides an opportunity to articulate repressed and forbidden life materials that resist, challenge or negotiate in some way with the status quo. The history of Israeli community-based theatre thus illustrates the complicated interaction between the establishment and the co-community in relation to articulation and empowerment. While community-based theatre is defined formally as a theatre created within, by, and for the 'people', in practice it is usually controlled and policed by institutional bodies.

Israeli community-based theatre, like that in America, is a form of 'believed-in theatre [...] in the fact that people are enacting their own stories and performing mostly for people of their own communities' (Schechner, 1997, p. 81). In Israel, it is the task of a professional director/facilitator to recruit and organize a group of local citizens into an ensemble. The creative process moves through several stages, always guided by the director/facilitator:

1 Developing a consolidated, creative group through exercises and games.

2 Acting out significant life events.
3 Using these personal experiences to form the basis for an original play collectively created by the director and the group.
4 Presenting a public performance of the resulting work mainly to a local audience.
5 Discussing the performances in a dialogue between performers and audience regarding the social intentions and messages of the performance.

Because the source of the community performance is not dramatic literature but 'community-owned' stories (Brady, 2000), the issue of authorship/ownership of the performance text is completely clear. It belongs to and is determined by the creative group itself. This issue-based, publicly performed 'self-text' expands the circuit of ownership to the entire local audience. However, this proclamation of the transfer of ownership of the theatrical 'means of production' to the 'people' is in practice merely a deceptive rhetoric concealing the operation of different regimes of power within community-based theatres. Community-based theatre in Israel has always been a function of the state's community planning and development programme. Community development, according to its own terminology, aims to bring about rehabilitation and progress within groups with special needs, such as ethnic minorities, at-risk youth, battered women, prisoners and the handicapped. Community-based theatre has become over the years a tool aimed at reducing cultural, social, educational and mental disadvantage within such groups.

Born in the 1970s in a few poor and disadvantaged neighbourhoods in Jerusalem and Tel Aviv inhabited mostly by Mizrahi Jews, Israel's community-based theatre is today found throughout the country in a variety of places, such as community centres, boarding schools, prisons and rehabilitation centres, engaging with different co-communities, such as Mizrahi Jews, Ethiopian Jews, Jews from the former USSR, Palestinians, disabled groups, prisoners and battered women. However, the Mizrahim still constitute the majority of community-based theatre performers.

Since its inception, community-based theatre has operated conditionally as a result of ideological disagreements between the sponsors and theatres concerning the meaning of 'change'. Does change involve the integration of the co-community into the dominant socio-cultural order or can it include resistance to integration? And if the latter, can the co-community make do with theatre as a symbolic weapon? The struggle over the meaning of performances constitutes a powerful manifestation of a complex politics of change, indicating that despite the rhetoric, co-communities never fully own the theatrical means of production. The transfer by the institutional bodies of the skills necessary for communities to make do with theatre is always partial and depends upon whether the production reflects or contests the state's hegemonic ideology. Therefore, the crucial and permanently relevant questions in community-based theatre are: who bears responsibility for the content of the event? Is it the municipal and state establishment that sponsors the theatre, the facilitator/director who as a theatre practitioner is often an outsider to the group paid by the establishment, or the creative ensemble that is supposed to compose a performance text important to the entire community? And what if the local audience that watches the show objects to all or some of the performance?

Politics of change within Israeli community-based theatre

The story of Israeli community-based theatre usually begins with the performance of *Joseph Goes Down to Katamon* (1972/73), directed by Arie Itzhak and produced by the Community Theatre of Katamon, an underprivileged neighbourhood in Jerusalem. The performance used the biblical story of Joseph in a cynical and ironic fashion. It offered a stylized depiction of the harsh and helpless daily life of the local disadvantaged inhabitants who, as Mizrahi Jews, felt discriminated against and oppressed by the dominant Ashkenazim (Jews originating from Europe and America). Itzhak was at that time a celebrated professional actor who became a social activist for his own ethnic Mizrahi co-culture. He was deeply inspired by the radical ideas that had reached Israel from America, and had already begun to realize them with at-risk youth in a shelter in downtown Tel Aviv. 'One day', recalls Yamin Masica, one of the performers in *Joseph Goes Down to Katamon*, 'a man approached us in the street. He looked different. He said that he had come from Tel Aviv to do theatre with us' (Masica, 2001). Itzhak was the charismatic type of activist identified by

Gramsci (2004) as the organic intellectual who knows how to make do with theatre in order to raise consciousness among young people, and who instigated a creative process in which the performers became his faithful followers. 'For us he was a hero who came from "there" and led us to an underground activity', states Masica.

Itzhak's dedication and enthusiasm was rewarded by the 'boys from Katamon', who stayed with him even after others found the work process too demanding and dropped out. These boys had never dreamt that anybody would really appreciate them, or that theatre was anything other than 'rich people's stuff'. Itzhak was usually accompanied by Michael Pharan, an outreach worker who introduced him to the municipal authorities. These agencies quickly embraced Itzhak's method as a means of rehabilitating streetwise young people. However, in contrast to the liberal intentions of the authorities, the young performers, influenced by Itzhak's socio-artistic agenda, made do with theatre as a political tool to protest against their conditions and initiate social resistance. The performers wrote songs and scenes representing poverty, crime, housing congestion and drugs. Itzhak introduced the biblical, metaphoric frame, and he also edited and moulded the personal materials. The young performers were deeply impressed and excited by their director's artistic choices, even if they didn't always really understand them, as Masica recalls.

One of the scenes, showing the brutal rape of a young prisoner, provoked objections from a large part of the audience. They found it too painful to watch, as well as injurious to their neighbourhood's already dubious reputation. The municipal delegates, mostly the Ashkenazi and the ultra-Orthodox, who rejected the overall oppositional tone of the performance, took advantage of the local audience's agitated reaction and asked Itzhak to cut the scene. Although the performers supported Itzhak's opposition to changing the scene, after long arguments he gave in, realizing that the alternative was to close the show (Miller, 1975).

The Katamon case formally marked the beginning of community-based theatre in Israel both communicatively and ideologically. At that time Louis Miller, a community psychiatrist, one of the founders of community work in Israel and also the senior psychiatrist at the Ministry of Health, became acquainted with the Katamon group. He gave it official recognition as the prototype of community-based theatre. In a number of articles Miller described the performance and laid out the basic model for such a theatre (Miller, 1973, 1975, 1977).

To theatre facilitators the Katamon case is ideologically important because it shows how community-based theatre can rise from the grassroots. But this is only part of the story. Katamon's creative process was able to reach its final stage largely due to the exceptional management of Michael Pharan, the outreach worker. As a community worker in the Municipal Department for the Advancement of Youth, especially involved with street gangs, he was sent to Katamon 'to calm down the sizzling winds, and to prevent agitation and disorder. A community worker was expected to be highly loyal to the municipal viewpoints and not to become too involved with the community' (Pharan, 2003). Pharan believed, however, that there should be an alternative approach, and when he met Itzhak he was 'enlightened by him'. He joined the creative process, assisting the group as much as he could, providing them with rehearsal space and equipment. He did not interfere in the creative process because he identified with it, considering that 'this is exactly what is needed to sincerely open windows' (Pharan, 2003). Pharan was indeed a rare individual, a member of the establishment whose personal social vision was in fact anti-establishment. However, he was unable to prevent censorship of *Joseph Goes Down to Katamon* by the municipal delegates; they had expected a grateful celebration, not a fierce social critique. Using their authority as the owners of the theatre, they demanded that the rape scene be cut. Shortly after this intervention, Itzhak left the group. The promising rhetoric of community-based theatre as 'of the people, by the people, and for the people' had been betrayed.

The Katamon case illustrates the two poles of the contested formation of community-based theatre in Israel: community development versus alternative-radical theatre. The former represented a liberal, patronizing approach to the Mizrahi co-culture as the 'other' in need of improvement and aimed at integrating it into the hegemonic order; while the latter, based on a neo-Marxist ideology, defined theatre as a weapon of the people that could be used to change the dominant social order. While the institutional sponsoring bodies did, and still do, expect a compliant performance, local theatre groups tend towards the creation of self-representational texts that make publicly clear their otherwise invisible wounds and needs. Generally, the more resistant the performance, the more it exposes the ideological gap between the community-based theatre and the establishment, thus intensifying the struggle over who really owns the performance.

At the beginning of the 1970s the Panterim Hashchorim (Black Panther) movement led a strong social protest among young Mizrahi residents of several poor urban neighbourhoods.[1] Most of these young people, who had immigrated after the establishment of the state, did not possess the skills necessary to succeed in a modern, rapidly industrializing society. The Mizrahim were perceived by the ruling Ashkenazim as backward and primitive. Politically, they had very little influence. They were usually sent to development towns, agricultural settlements or designated areas in the larger cities, which quickly became slums. These newcomers became the Jewish proletariat, which was (and still is) largely connected to its ethnic origins. This correlation between ethnicity and class created a complex economic, political and cultural gap between the hegemonic Ashkenazim and the alienated Mizrahim.

For several years, the Black Panthers shook Israeli society with large-scale, violent and provocative demonstrations, focusing public attention on the poverty, deprivation and discrimination by which Mizrahi Jews were oppressed. This disturbed socio-political situation generated the radical community-based performances that showed the Mizrahim as resisting subjects. The struggle over ownership/authorship was won by the establishment, which eventually blocked the activity of community-based theatres.

In 1977, the right-wing Likud party came to power for the first time. The most striking feature of this election (and of the one that followed) was the support that Likud received from the Mizrahi co-culture, who blamed the Labour Party for years of discrimination.[2] This change in the political power base tempered the unrest among the Mizrahim who felt that the new Likud government would improve their conditions. This optimistic atmosphere determined the non-radical character of the second wave of community-based theatre which presented the Mizrahim as appeased, co-opted subjects. These kinds of performances celebrating the new status quo and creating the impression that the people owned the theatre were welcomed by the authorities.

Since the 1990s, as Israeli society has progressed toward more cultural openness and relativism, the ownership/authorship issue in community-based theatre has moved into its third phase. Theatre groups, now more conscious of the power of the sponsoring body, deliver their oppositional messages (when they exist at all) implicitly rather than directly. At the same time, the establishment, which is increasingly aware of the politics of multiculturalism and political correctness, now uses more sophisticated strategies to control the performances. The creative process is usually autonomous, and it is only close to and/or after the first performance, that explicit manifestations of the struggle erupt.

1974: *The Other Half*

The radical text of *The Other Half* (1974) was generated within the context of a powerful fusion between the contemporary national socio-political reality – the protest of the Black Panthers – and a particular local socio-political and economic reality. Pardess Katz is an underprivileged Mizrahi neighbourhood in the municipality of Bnei-Brak in Greater Tel Aviv, an ultra-Orthodox Ashkenazi town occupying both sides of a large inner-city road that geographically, economically and culturally divides the dominant Ashkenazi half from the Mizrahi 'other' half of the population.

Yossi Alfi, a professional actor who had returned to Israel from London after two years of theatre training at the London Academy of Music and Dramatic Art (LAMDA), where he was influenced by English alternative and community-based theatre, was one of the founders of Israeli community-based theatre. After working with prisoners and students, he sought out a disadvantaged neighbourhood in which to organize a theatre group. Arriving in Pardess Katz in 1974, he persuaded the local community worker and a few important municipal members to support a theatre. Alfi's objective had been to work with young unemployed dropouts. He offered them the theatre as a platform from which to express themselves and tell the story of their community. As a prerequisite for the project, Alfi asked the Pardess Katz inhabitants questions about their daily life and problems in order to become acquainted with the neighbourhood. Alfi was mistrusted from the beginning by the local politicians, who had been elected by the ultra-Orthodox Ashkenazi municipality and the Labour Party, and entrenched local leaders reacted with great anxiety, trying to prevent him from developing too deep a relationship with the inhabitants. They offered him a very low salary, hoping he would give up, but dedicated as he was to the idea of community-based theatre, he decided to stay (Alfi, 1986). Alfi wanted a community worker to be his partner, as in Katamon, but each worker was in turn 'accused of inciting the inhabitants' and was fired (94). Despite the problems, Alfi managed to organize a group of around twenty teenage boys and girls.

Working for about a month, Alfi used a tape recorder to gather information about the daily life and problems of the young people of the area. He and the group also met their neighbours and recorded their problems, dreams and complaints. Next, Alfi formed the material into a play which he reworked with the group. The play described the difficult conditions of life in the neighbourhood, its shabby appearance, and the alienated and harsh administration of the ultra-Orthodox bureaucrats. In one of the scenes Orit talks about her mother:

> My mother used to tell me that once, when they came to Pardess Katz, it was bad. Now it's no longer bad, it's worse. She speaks about the transit camp as if it was paradise compared with what is happening now. Once, she could dream. Now my mother's dreams are stopped by our tiny apartment's concrete, ugly walls.[3]

In another scene, Mister Katz beats his dog, Pardess, who wants to eat the meat promised to him. Mister Katz, the master, shouts:

> Come here! Don't move! Sit! Only trouble I have from you, only trouble, you primitive! And after all the good I have done for you [...]. I saved you from all the Arabs, I brought you here in an airplane, and did you ever fly before? Since when do dogs eat like people? [...] When the Messiah comes all the creatures will be equal. Until then, sit and be quiet.

The authorities were highly suspicious of Alfi's method of playwriting, charging that he was deliberately targeting local social problems and interfering with their own work. 'They saw in me a "snooper"' Alfi stated, 'the claim against me was "you are a director, so direct and that's that. What are you looking for all the time, who said what to whom, who did what to whom?"' (Alfi, 1986, p. 94). The authorities wanted Alfi to direct the kind of play that would get the youth

off the streets while entertaining the locals. They vigorously objected
to any performance that criticized their own conduct.

But the script indeed 'aroused problems', affirms Miriam Huppart,
a high-ranking social worker in the welfare department of the
municipality: 'The material described the neglect in the neighborhood, the severe failures of the establishment, especially in relation
to the activities of the local municipality, and the deprivation of the
neighborhood in contrast to other neighborhoods in the same city'
(Huppart, 1975, p. 30). According to Huppart, the script, which was
submitted for approval before it could be produced, provoked: 'a dis-
putation with the establishment's delegates [...]. It was difficult [for
them] to digest the critical and aggressive text [...]. It created a feeling
of uncertainty concerning the control over the aims and contents of
the theatre' (30).

Nevertheless, after much argument, in which Alfi, appropriating
poaching (subversive) tactics, emphasized the comic nature of the play,
the authorities approved it. *The Other Half* opened at the beginning of
June 1975 in the local community club in front of local residents,
the municipal delegates and a few journalists. The show consisted of
musical pieces and short sketches that delivered sharp, oppositional
messages directly accusing the local ultra-Orthodox municipality:

> They split the town into inferior and superior with one road
> between the two halves. They enrich the former while for us they
> only bring frayed nerves.

> I have nothing against religious people [...]. I myself believe in
> God. But I have a lot against those religious politicians in the
> municipality, who cook up our life here.

The performance pointed out how the local municipality represented
a painful microcosm of national state policy:

> Who am I in this country? [...] I'm the second Israel. I want to be
> a first-rate citizen in this country. Why do the big shots ask me
> to step aside? [...] I'm the black sheep of the country. The differ-
> ence between me and an Arab is only one small grade. They [the
> establishment] want us to remain nothing, they look at us as if we
> are animals, savages, and primitives. They think for me and decide

for me. They crushed my own culture and made me a scarecrow.
So what is left for me? To scream and shout till my last day, Stop
stepping on my soul and throat!!!

The Arabs are mentioned briefly and disparagingly here for the only
time in the production. At that time, the Mizrahim were preoccupied
with making a living and used community-based theatre to fight
primarily for the benefit of their own minority. In *The Other Half*
the Mizrahim resisted their social and economic proximity to Israeli
Arabs because they themselves were only a little more privileged.
These two co-communities were fighting each other for the same
small set of resources. Mizrahi Jews during the first two decades of
the State of Israel were forced to repudiate their Arab background in
order to become a part of Israeli society. Their similarity in appearance and culture to the Arabs threatened the hegemonic Zionist
project that defined Arabs as the enemy 'other'. The working-class
Mizrahim were generally in line with the Zionist-Ashkanazi ethos in
negating the Arabs.[4]

It was not only *The Other Half*'s text that was radical, but the staging as well. The production replicated a political revue that Alfi had
imported from the English working-class theatre tradition and the
alternative theatre of the early 1970s. *The Other Half*'s poaching (subversive) poetics appropriated, for example, popular melodies from
Hebrew pop music as well as Jewish liturgical music. Moreover, the
two central images were that of the Mizrahi as a scarecrow and as a
dog, representations which powerfully articulated the low cultural
and economic status of the Mizrahim as a consequence of Ashkenazi
discrimination. The stage was open and empty, emphasizing the
plainness of the neglected local public hall. There were only a few
posters on the backdrop and some placards and props on the stage.
In the dialogue and songs, Hebrew was combined with Iraqi Arabic,
which was the mother tongue of the older generation in general, and
the performers' parents in particular. Yiddish words and expressions
were also scattered in the text and their exaggerated pronunciation
by the Mizrahi performers parodied the Ashkenazi ultra-Orthodox
Jews while at the same time erasing the cultural differences between
Ashkenazim and Mizrahim, as the Mizrahi performers gave a highly
successful impersonation of the ultra-Orthodox. The atmosphere of
resistance was enhanced by the rapid flow of brief, sharp episodes,

performed in a direct, rough, rhythmic and enthusiastic style with group bodywork. Every scenic choice was made to reinforce the text, and thus posited an example of grounded aesthetics – that instrumental and activist local aesthetics which fuses the politics and the poetics of representation.

The Other Half was too much for the local municipal leaders. 'The institutions that were responsible for the theatre could not accept the idea that they could not intervene' (Huppart, 1975, p. 29). After a few performances, Alfi reported, 'the struggle to remove it from the stage has begun; the authorities found it insulting' (Alfi, 1986, p. 96). Alfi was ordered to make major changes in the text. When he and the group refused, the authorities immediately and angrily dismissed him and disbanded the theatre. Thereby proving that the sponsoring body had never intended the community theatre to have either artistic or political autonomy.

1982–85: *Waiting for a Saviour*

In 1982 a community theatre in Ramat-Amidar (a disadvantaged neighbourhood in Ramat Gan near Tel Aviv) was revived as part of 'Project Renewal', a plan instituted by the new right-wing government, financed mainly by Jewish donations from the Diaspora. The project aimed to allocate large budgets to disadvantaged areas throughout the country for the physical renovation of dilapidated apartment buildings and the cultural integration of the people who lived in them. In each neighbourhood, a board consisting of central government delegates and local representatives was organized to decide how best to use the money. The attachment of community-based theatre to this project was based on its educational and rehabilitative potential. This 'renovation' policy for both buildings and inhabitants was in fact only a variation of the familiar hegemonic approach of the establishment to community-based theatre. Nevertheless, the new cooler temper of the Mizrahi inhabitants themselves, who hoped that the new Likud government would offer a different policy from that of the Labour Party, led them to welcome 'Project Renewal'.

Yossi Alfi, after his difficult experience in Pardess Katz, modified his vision, recognizing that although community-based theatre should not ignore politics and claims for change, theatre activism should be based on participation in the political order not protest against it

(Alfi, 1986, p. 78). Together with students from the Community Theatre Unit of Tel Aviv University, he established the Ramat Amidar theatre, which became the leading community-based theatre of the 1980s. Three stages were planned for the three-year project: (1) the organization of theatre projects in which various groups worked autonomously; (2) theatre training; and (3) training local staff to lead projects independently. Ramat Amidar opened its doors to the whole community, encouraging not only street youth but all the neighbourhood's children, young people, adults and senior citizens. Theatre activity took place in elementary schools, senior citizens' clubs and the local cultural centre.

Alfi's newly liberal agenda suited the temper of the local theatre groups better than the earlier radical approach. The performances during the three years were entertaining in style and dealt with universal themes, such as love, success and interpersonal relationships (Zeltzer, 1986). Because they were integrative and not confrontational, the first year's performances were more like festivals than protest theatre.

The first production, *Neighbourhood Day*, took place during the Jewish holiday of Lag Ba'Omer when people traditionally light fires to celebrate the victory of Bar Kochva, the Jewish warrior who defeated the Romans in 132 CE. Alfi organized a colourful festival, beginning with a participatory parade featuring the town police band, firemen and their red engines, and children from the local kindergartens, schools and youth organizations as well as groups from the community-based theatres. The central event was staged in the local park where members of the municipality congratulated and praised the community theatre. This was followed by the presentation of several dramatic pieces performed by a number of the theatre's actors.

The second year's event, *The End of the Year Project*, was staged indoors at the local cultural centre. All the dramatic productions of the various community-based theatre groups were shown. Together, the pieces comprised about an hour and a half of low-quality performance based on improvisation and role-playing. One of the better pieces was *The Miraculous Island*, written and performed by the senior citizens' group. The plot followed a journey to a secret island where the elderly people found a drink that transformed them once again into young people. This performance was warmly welcomed by the audience. Discussing American community-based theatre, Richard Schechner distinguishes between two forms: the first is 'excoriating', which represents 'avant-garde and disruptive voices', and the second is 'celebrating', which honours the dominant culture and 'celebrates the community, even when painful' (Schechner, 1997, pp. 80–1). The first and second years at Ramat Amidar exemplified the celebratory mode of community-based theatre.

In the third year the two most prominent productions were *Waiting for a Saviour*, which was scripted and performed by the youth group, and Molière's *George Dandin*, staged by the adult group. The director/facilitator of the youth group was Igal Azarati, then a graduate student and today a leading political theatre practitioner and the manager of the Arabic-Hebrew theatre in Jaffa. Azarati recalls that he came to the project very committed to activist community-based theatre. He was astonished to find a group of youngsters so deeply self-involved, so different from the radical youngsters of the 1970s. However, if he had thought that in time they would become more politically aware, he was disappointed (Azarati, 2001). Despite his misgivings, he followed the self-centred agenda of the youngsters because he thought:

> A community-based theatre director should do what the group wants, and not realize his own viewpoint. At that time I had been sentenced to prison as I had refused to do military service in the occupied territories. The group knew my political views, and even came to demonstrate in front of the jail, but in their own theatre they didn't see any interest in discussing the politics of their own ethnic identity.
>
> (Azarati, 2003)

The creative process was based mainly on improvisation. Azarati, together with the performers, scripted the successful exercises into a play. *Waiting for a Saviour* parodied the excoriating community-based theatre and the ideology of community-owned stories. The plot concerned a group of young community-based theatre performers playing themselves on the stage, waiting for their director to come and save their 'ridiculous, oversimplified, melodramatic, and in short, rubbish show'.[5] In the meantime they try, not very successfully, to mould a few typical, local scenes into what they cynically define as 'a serious and important show that will shake the neighborhood and push the people out into the streets! This is what we have a theatre for, isn't it? To agitate them! To throw the truth in their face.' However, the real performers themselves appeared to have had no interest at all in such a theatre, depicting local social issues such as alcoholism, drugs, violence and prostitution in an exaggerated and stereotyped manner. In the play, the fictional director, played by Azarati himself, eventually arrives, only to announce he is leaving for good: 'I deliberately didn't come. I'm fed up with you, and have no more energy for you. You can manage perfectly well without me.' If this statement represented Azarati's true reservations regarding such a community-based theatre, none of the group, the audience, or the municipal delegates appeared to have caught on. Everyone, locals and authorities alike, enthusiastically applauded the performance (Azarati, 2001).

In *George Dandin*, the group chose to abandon the basic model of 'community-owned' stories, mounting instead a classical play. In fact, this step completely changed the theatre from a community-based theatre into an amateur theatre, which, as Schechner indicates, is the very opposite of community-based theatre because it no longer shapes local materials but stages popular hits and classics (Schechner, 1997). The sponsoring authorities clearly preferred an amateur community theatre to a disruptive community-based theatre. There were no conflicts over the text or the production. Everyone was happy because the aesthetics were familiar and there was no political challenge.

1996–97: *Phachme Dast*

Over time, the Mizrahim have gained increasing political representation. There have been signs of a willingness to include new voices and once again seriously to discuss the sharp divides in Israeli society. In the late 1990s, in Neve Eliezer, an underprivileged neighbourhood of Tel Aviv, director Peter Harris staged *Phachme Dast* (an Arabic expression meaning 'dust and ashes'). It was outwardly more promising than the tame performances and projects of the previous two decades. Community-based theatre in general, and in Neve Eliezer in particular, won support and recognition from the establishment. Nevertheless, or maybe because of this, *Phachme Dast* was not a celebratory performance: its messages were even more radical than those of *The Other Half*. In the 1990s, more aware of the dilemma concerning who owns

Israeli culture, the director/group and the authorities strove to handle the question in a more sophisticated and subtle manner.

Harris had emigrated from England in 1976 as an expert in youth and community studies, and after completing his professional and academic theatre training at Tel Aviv University, he went on to become one of the leading figures in community-based theatre. The play's narrative concerns the hidden history of the neighbourhood from the early 1950s, through the 1970s to the late 1990s. It is a history of crime, alcoholism, violence and religious fanaticism, told through the life stories of several characters, among them a local crime boss, a political activist, a strong woman and a desperate musician. Harris, for whom 'the community performer is the central artist and the owner of the text', preferred the workshop/collective creation method (Harris, 2003).

The eighteen performers represented the various ethnic groups living in the neighbourhood: Jews from Egypt, Yemen, Morocco, Iraq and Iran. Harris met the group twice weekly, devoting one meeting to theatrical practice and the other to playwriting. The performers brought their own stories about the neighbourhood from the three different time periods. The older Yemenite immigrants revealed the history of the first years, the 1950s: how they had inhabited a deserted Arab village where they enjoyed a rural lifestyle, and how they were tempted by the promises of the authorities – dreams of modernity and a better education for their children – to move to small, crowded apartments. The Moroccan and Iranian immigrants of the 1960s and 1970s recalled their arrival in the slums and the gang fights between the different ethnic groups. The youngest performers told how Neve Eliezer had become a centre for drugs and crime, and how many of the inhabitants escaped their miseries through addiction to drugs, alcohol, gambling and fanatical religion.

Harris tape-recorded the stories, and then added one contemporary scene. This episode focused on a hostile municipal representative, whom the performers firmly demanded should be represented as a Mizrahi local inhabitant. 'For us', noted Meyira Medina, one of the leading performers and the director's assistant and today a community-based theatre facilitator, 'the worst is when a Mizrahi "goes Ashkenazi", and then treats us even worse than the Ashkenazim' (Medina, 2003). This scene became the dramatic hook for the play, which was scripted in a workshop process by five of the performers and Harris. The final script was created through discussions with the other members of the ensemble.

'In 1997, the Jubilee year of the State', noted Harris, 'we thought it important to represent 50 years of oppression, and deliver the message that something has to be done, and quickly' (Harris, 2003). *Phachme Dast* posits a counter-memory to the national Zionist master narrative and its glorified descriptions of the heroic and courageous foundation and development of settlements celebrating the Zionist ethos. The story of a neglected area, mostly inhabited by Mizrahi immigrants, contradicts the official rhetoric of equality, justice and cultural integration. Such a story therefore was, and still is, excluded from formal histories. *Phachme Dast* articulated the hidden history and memory of a Mizrahi co-community, but at the same time it continued to exclude another repressed memory – that of the Arab community that had once lived there. The questions of why and how this place was deserted when the Jewish immigrants arrived were never raised. The Mizrahi co-community identifies with the well-defined borders of the Jewish national collective and strives to distance itself as far as possible from the Arabs in general, who are perceived as the essential enemy, and from the Israeli-Palestinians in particular, who are considered the lowest stratum of Israeli society.

Nevertheless, this forgotten, unwanted and concealed historical narrative did slip into the performance, as if unconsciously signalling that the mutual negation of the Palestinian and Israeli narratives is never total: the absence of one narrative is somehow always traceable in the representation of the other. In *Phachme Dast* the nickname 'village', which the inhabitants of Neve Eliezer still use for their neighbourhood after so many years, is spoken onstage and was also indicated in the scenery. The central visual element comprised six large grey columns – an indexical sign of the ugly apartment buildings that typify the neighbourhood. In the scenes that represented the 1950s, these columns, which were painted on the reverse side as blocks, were turned around and connected to create the appearance of an old wall of broken bricks. In a scene that critiqued the Ashkenazi inhabitants who had moved on to more affluent suburbs the moment they received their German reparation money, Kopel, a Holocaust survivor, excused himself to his Mizrahi neighbour. He explained that as an immigrant from Europe he could not live in such conditions, in an abandoned house, which might be good enough for an Arab or a Jew who came from an Arab country, but not for him. Totally immersed in their own misery, the Neve Eliezer community-based theatre failed to grasp this opportunity to confront the issue of silenced history, which is inscribed in the evidence of the ruined walls of a mosque that still stands in the neighbourhood.

While *Phachme Dast* warmly recalled the Arabic cultural roots of its community by representing onstage the music of Umm Kultum, the famous Egyptian female singer, and various traditional Arab practices against the 'evil eye', the dissociation of the Mizrahi community-based theatre from the Palestinian-Israeli conflict was evident throughout the play.

One of the characteristics of the 1990s was the legitimization of Mizrahi ethnic culture in the centre of Israeli popular culture. But even this did not affect the negation of the Arabs by the Mizrahi working class. On the contrary, this Jewish co-community is still fighting forcefully to differentiate itself from the Arabs. This of course indicates that a co-community is not necessarily empathetic or free of bias in relation to another co-community. When a co-community such as the Mizrahi makes do with community-based theatre, it is first and foremost for its own self-constructed needs. However, there is a small group of Mizrahi critical scholars and artists that has become acquainted with post-colonial thought and action. This group has recently been trying to deconstruct the Zionist discourse by affirming the link between the Mizrahi and Arabic cultures and by redefining the Mizrahim as Jewish-Arabs.[6]

The narrative of *Phachme Dast* was violent. It carried a threat of social unrest and cultural separatism. For example, Murduch, one of the older, critical characters of the play, recalls his mother's joy when she entered the new apartment the authorities had given her. While the memory is reconstructed through role playing between mother and son, Murduch jumps from the represented context of the 1950s to real time, turning directly to the audience:

These cursed matchboxes, how could we raise children when five or eight of them were sleeping in the same tiny room? People who live in matchboxes will eventually burn, and the fire will light a fire, and there will be a huge conflagration.[7]

A second example is the scene in which Esther encourages her brother Shalom to be optimistic and to try to find a way out of their harsh life. He answers: 'They shoved us all into a 70-square-meter apartment, a matchbox, and bye-bye. The State did the minimum for us and now would like to send us – her niggers – to hell. [...] We will take our revenge on them.' When Shalom realizes that his problem is not individual but characterizes his entire co-community, he becomes a social activist, and in his speech in the last scene – both to the fictional community onstage and to the real community in the auditorium – he proclaims: 'The time has come for us to stand up, all of us, and take our basic rights [...]. The time has come to take action [...].'

Another example is from the scene that recalls a music lesson in which the Ashkenazi teacher humiliates the Oriental music of her young Mizrahi pupil. Zion, the pupil, reacts angrily: 'Who are you to judge my culture as primitive? [...] I really don't see a possibility of bridging between your culture and our garbage. So let's remain each with his own roots.'

In contrast to his radical predecessors of the 1970s, Harris was well aware that he was under the surveillance of the authorities, who effectively held the fate of the theatre in their hands. He therefore acted as a poetic poacher and strove to achieve a balanced dramatic rhetoric that would operate as a marker and a mask at the same time. The radical text was organized in such a way as to confront the Ashkenazim at large, and mainly for their deeds in the past. This carefully balanced approach was echoed in the scenery and staging which marked the historical time and place of each scene. The permanent backdrop was a large abstract painting that vaguely presented 'a neighbourhood' under a bright sky. The moveable scenery, as mentioned before, comprised six columns on wheels that were painted on one side in gray, to indicate the concrete pillars of public housing, and on the other side as bricks with moss, to signify the old Arab village. The indoor scenes were indicated by props such as chairs, a table, paintings and clothing. In the left corner, near the musician, stood a placard displaying the particular year, which was changed by the musician for each scene. The acting style, which was realistic in general, was intercut with original songs which had pleasant melodies and words that made no particular reference to the neighbourhood. The radical messages were, on the one hand, here and now, but on the other hand, they were properly marked as recollections of the past. The only radical scene that represented the present time was the demonstration and the speech by Shalom at the end of the performance – and this indeed became the only problematic scene, both in its process of creation and in its reception.

Another tactic was self-reference, which the group used in one of the scenes toward the end of the play. When the fictional Mizrahi head of the educational unit in the municipality tries to justify his attempt to get rid of the neighbourhood representative with the comment that 'We gave you people everything, even a theatre', he is answered with: 'Yes, in order to shut our mouths, so that we will appreciate you for letting us express ourselves [...], you'll probably close the theatre when you hear what we have to say.' By presenting their objections as part of the performance itself, the performers hoped to forestall the habitual response of the sponsors whenever community-based theatre became too political. 'I wanted the municipal representatives attending the performance to be affected by it', said Harris, 'but not to the extent that they would walk out of the auditorium' (Harris, 2003).

Nevertheless, Harris had to fight against the head of the municipal drama department, who as Harris recalls, 'saw himself as a kind of a dramaturg and demanded to read, respond to, and approve the text' (ibid.) After watching the dress rehearsal, he was furious, demanding that the demonstration scene be cut. He probably objected because the scene was too provocative and politically dangerous, but he argued that his sole reason was artistic: one should retain the 'fourth wall' to the end, for fear of offending the audience. Harris strongly opposed this demand, and responded in the same vein. For him and the group it was artistically important to involve the audience actively in the last scene. The head of the municipal drama department, according to Harris, even threatened to cancel the show and bar the group from the annual community-theatre festival.

Harris, however, backed by the group and the local residents, did not give in. Finally, the scene was allowed. This incident reveals the new 'soft' mode of rhetoric that the sponsoring bodies employ in order to control the production. Trying to argue their position on artistic grounds is another political strategy that appears in a professional disguise. Nevertheless, it seems that Harris caught the sponsoring body's real intention and was able to counter and to gain the final victory by using the same rhetoric.

In fact, the director and the group had themselves already fought over this scene during the work process. The group wanted to end the performance with the clearly aggressive act of burning down the old youth club, which had become a drug site, but Harris finally used his authority and ended the play more vaguely, emphasizing an 'artistic' argument to justify his decision. In the mass demonstration in which Shalom, the peaceful political worker, changes his conciliatory, co-opted attitude and declaims an oppositional speech, a mysterious fire breaks out in the background. In David Ophek's documentary film (Ophek, 1997), which followed the creative processes/rehearsals and the premier, I found a brief fragment in which the performer who played Shalom insisted that he should light the fire in front of the audience. Medina recalls the incident:

> He was really very nervous. He fought Peter [Harris] strongly, and I think he was right. He thought that Shalom represented most his neighborhood and if the municipality would not provide a proper decent way of life for the children, he would have no choice but to join the militants.
>
> (Medina, 2003)

In the film Harris is seen to argue against the idea, saying: 'Let's check it artistically.' When I asked him about it he insisted that the lighting of a fire by Shalom, a reformed criminal, was a bad artistic choice (Harris, 2003). In this particular conflict, the final decision was the director's. Medina, who, besides performing and scripting, was also the director's assistant, interprets the position of the director in the community theatre very clearly:

> Peter was stuck in the middle, between us, who were very asser-tive, and the manager of the community center, the head of the drama department, the community workers, and other municipal delegates, who were watching him. He had to satisfy them too.
>
> (Medina, 2003)

Phachme Dast won great acclaim, and was performed several times in front of packed audiences. In the discussion between the thea-tre group and the audience after the performance I attended, the atmosphere was one of great approval. The only reservation, one of the municipal delegates said in a humorous and friendly tone, was that he hoped in the next performance the text would reveal more accurately the many 'good things' the municipality had achieved in the neighbourhood 'throughout the years'.

Naftali Shem-Tov, one of the performers and script-writers and today an academic theatre scholar, recalls that the apparently good feelings had led to glossing over the strategies that some of the municipal representatives had carried out in order to control the text: 'I was one of the five performers who were involved in the play scripting. It was I who wrote Shalom's radical speech in the demonstration' (Shem-Tov, 2002). The speech alludes to the oratory of Martin Luther King at the beginning and to the concerns of the Mizrahi Black Panthers at the end:

> I have a dream... One day this neighborhood will be a model for the whole city. You, the people who are standing in front of me, will be happy and proud of yourselves. I have a dream... One day all the tags attached to us will vanish, and our ethnic and cultural origins will become a source of self-esteem for us and for our children. [...] We are sick of all the promises of the politicians. [...] They did us a favor and stuck us into matchboxes. [...] They decided upon renewal, Project Renewal. [...] They paint the ghetto walls, they brought social workers to define us, to examine our children in order to prove that we are inferior. [...] We will stand up together and reclaim our basic rights, the right for proper education, the right to proper living, the right to be Mizrahim [...].

After the first performance, a high-ranking official of the welfare department became upset and annoyed. Shem-Tov observes that:

> She was afraid, I think, that this monologue would really goad people in the audience into taking steps. I think she was worried about her own position. She didn't want any trouble. She didn't do anything directly or aggressively, as in the old days of community-based theatre, but nevertheless, she invited us to her nice home for an 'intimate encounter,' and there, along with another big shot from the municipality, they asked us very diplomatically whether we had any operational plans to encourage real protest in the streets, or become involved in the local elections. Most of the people chickened out. 'We are small people, we have no political

power,' they said. This wasn't a tactic but a genuine expression of how they still really felt.

(Shem-Tov, 2002)

Initially, according to Shem-Tov, the concluding monologue had been even more extreme. After the first performance, however, Harris told the group that some municipal representatives thought that the sentence, 'Enough, we will not be black cockroaches of the Ashkenazi establishment', went too far. 'I didn't think so', says Shem-Tov, but he eventually followed the director's instruction and cut out the 'bad' line.

'Theatre of the people': nonetheless, making do with theatre

Discussing here the issue of ownership in Israeli community-based theatre as characterizing the complexity of making do with theatre, I have demonstrated the shifting dynamics between the radical and the co-opted approach to the written and performed text in the struggle between the community-based theatre and institutional forces. The promise to provide a voice to co-communities has been conditional on consent and gratitude. From the early radical experiments of the 1970s, to the celebratory events of the 1980s, to the more recent subversive performances, community-based theatre continues to operate between the Israeli authorities and the community authors. Today, when both the theatre groups and the institutional bodies are more aware of and experienced in the complicated power relationship in which they are involved, they strive to handle the tensions more reasonably and less abruptly. Although the co-community still only partially owns the production process, and is persistently warned about becoming too contentious, it nevertheless, eventually, makes do with theatre. The examples I have discussed here, mainly of more radical and subversive theatrical events, demonstrate how by using poaching poetical tactics the performers succeeded in mounting a grounded aesthetic performance that publicly articulated the resistant voice of their co-community loud and clear.

6
Battered Women on the Stage: from Spoken Objects to Speaking Subjects

If we shift our focus of attention away from the mainstream theatre, where crucial feminist issues are in general carefully avoided, we can locate those theatre sites where women make do with theatre for their own needs. This chapter focuses on two theatrical events in which women articulated the issue of violence against women in very different community-based performances.

Despite statistics indicating that violence against women in Israel is steadily increasing and that 15 per cent of all families currently suffer domestic violence,[1] this crucial subject, which is finally visible and indeed something of a preoccupation in the mass media, has not yet been symbolically confronted by the country's major theatres. However, in contrast to its invisibility on mainstream stages, violence against women has been the central theme of several non-institutional performances, including *Battered Women* (1981) by the Neve Zedek Theatre Group,[2] *Every Seventh Woman* (1997–98) by the Shkunat Hatikva Community Theatre (located in an underprivileged neighbourhood in Tel Aviv inhabited mainly by Mizrahi Jews), *And That God Will Help You* (1999) by the Jaffa D' Community Theatre (also in Tel Aviv, in an area populated by Israeli Arabs and Mizrahi Jews), and *A Plague not Written in the Bible* (2000) by the Centre for Prevention and Treatment of Domestic Violence in Herzlia, in central Israel.

Battered Women was a documentary theatre piece, directed by Nola Chilton[3] and performed by professional actresses who had just graduated from university, and who were mostly of Ashkenazi origin. *Every Seventh Woman* and *And That God Will Help You* were produced by amateur actresses of Mizrahi background, who although being acquainted with domestic violence had strongly avoided being labelled as battered women. *A Plague not Written in the Bible* was performed by a group of amateur actresses, whose low income neighborhood, unlike the previous two groups, was a heterogeneous mix with a Mizrahi majority, and who were all being treated at the Centre for the Prevention and Treatment of Domestic Violence. In the first production, the character of the battered woman was spoken by another, culturally-distant woman; in the following two productions, she was presented by a more closely identifiable relative; but only in the fourth production was she publicly represented by herself, articulating her own voice and signifying herself through the symbolic medium of theatre.

This chapter discusses the first and the last of these four productions as they most clearly show the similarities as well as the differences between documentary theatre and community-based theatre. The focus here is thus on the process of passage from acting woman-as-object to acting woman-as-subject, from being made by theatre to making do with theatre.

Battered Women: documentary theatre as social activism

Battered Women was co-produced by Tel Aviv University and the Neve Zedek Theatre Group in 1981 in Tel Aviv. To appreciate the significance of this performance as a cultural text requires some understanding of the issue of battered women as perceived within Israeli public discourse at the time. At the beginning of the 1980s the first three shelters for abused women in Israel had just been founded. An investigation by the Interior Committee of the Knesset (Israeli Parliament) had determined that out of 1500 women who had sought legal/political help, 55 per cent testified to having been battered. The committee reached the conclusion that although battered women generally tended to be silent about the violence that they suffered, only 5–10 per cent of all married women in Israel were being abused (Svirski, 1978).

The general perception of violence against women was that it was an esoteric, ethnic phenomenon to be found in the low-status Mizrahi community. Within the dominant Ashkenazi discourse, battering women was seen simply as another sign of the backwardness of the Oriental subculture, which 'permitted' the Jewish or Arabic husband to batter his wife. This interpretation served further to legitimize the establishment's 're-education' projects, which sought to transform the Mizrahi co-culture from the 'Other' to part of 'Us' as soon as possible. Because domestic violence was defined as an extension of an ethnic problem, there was no policy specifically addressing the problem of battered women.[4] This patronizing interpretation silenced discussion and excluded the issue of battered women from the public sphere. Against the backdrop of this socio-cultural invisibility, *Battered Women* represented a primary symbolic disruption of the status quo, especially in its implied call for social agitation.

The activist impulse of the performance was directly linked to the artistic agenda of Nola Chilton, a leading theatre director and teacher. Chilton was born in the USA, where she became involved with experimental theatre for social change and acquired a name as an important director and acting coach both on and off Broadway. When she immigrated to Israel at the beginning of the 1960s, she brought with her the genre of documentary theatre and the American social activist spirit, along with an intention to introduce them both to the Israeli social context. Her main artistic base between 1970 and 1990 was Haifa Municipal Theatre, which at the time was seeking socio-artistic distinction as a leading theatre operating at a geographic and intellectual distance from the cultural centre of Tel Aviv. Chilton took the theatre in new directions and her artistic work has continued to contribute to Haifa Municipal Theatre's centrality on the socio-political-artistic map. Through productions such as *Joker* (1975), *Crisa* (1976), *A Bicycle for a Year* (1977) and *Endgame* (1978) (Urian, 2000), Chilton became the first to confront the ethnic problem on the established stage. She challenged the actors (mostly of Ashkenazi origins) as well as the theatregoers (also mostly of Ashkenazi origins) by introducing Mizrahi characters as protagonists, thus marking as an issue that which was constantly being erased as a non-issue.

Chilton introduced community-based methods, urging the actors to live within the co-community, each carefully studying one individual as a source for a prototype character. To reinforce the documentary level of the performance as an accurate citation of the social reality, she used a hyper-realistic acting style and directed the actors to avoid 'pretence', thereby minimizing as much as possible the 'as-if' principle. She saw herself as the creator of a theatre that provided 'a rare opportunity to people on the fringe to be in the center', and was sincerely determined to use socio-documentary plays 'to penetrate and change the process', which she described, following Foucault, as 'abstract awareness, ignoring the human reality, which frequently accompanies denial of the "Other"' (Shapran, 1978). This activist ideology and practice also underlay the *Battered Women* project, which nevertheless focused on the problem of battering not as an issue in itself but rather as another symptom of the Mizrahi ethnic problem.

The artistic process

A few years prior to the production of *Battered Women*, around the time that the first shelters for abused women in Israel were opened, Chilton had turned down a request from a shelter to produce a show on this sensitive topic. This *'pasionaria* of the deprived groups' rejected the request on the basis that 'the feminist issue is not something that turns me on [...] there are more urgent subjects in the country [...] I am simply against feminist propaganda' (Naaman, 1981). However, the idea came up again, this time from a very different source. As an acting teacher at the theatre department of Tel Aviv University, Chilton was asked to produce a documentary performance with the graduating acting class, which uncharacteristically that year consisted solely of women. Among the various possible issues for treatment, she was approached by one of her colleagues with the subject of 'battered women'.

Over a period of eight months, seven students visited the shelter for abused women in Herzlia, each forming an intimate relationship with one of the women there, listening to and recording her story, and adopting in detail her way of speech and movement. These taped materials were edited, arranged and directed by Chilton to appear as a 'literal transfer of raw material to the stage' acted out 'as a piece of social action, making people aware of a sad phenomenon in our society' (Kohansky, 1981).

Thus, her decision to deal with the issue of battered women was primarily one of functional necessity, and its relevance lay in the potential for appropriation of the subject as an additional case study for her theatrical representation of the ethnic problem. This position, as I intend to show, determined all Chilton's poetic strategies, which ultimately objectified the battered woman and, as such, perversely sensationalized her case.

Performance analysis

After several performances at the University Theatre of Tel Aviv, *Battered Women* moved to its new quarters at the Theatre Group in the then slum neighbourhood of Neve Zedek. This new marginal geographic setting acted as an extension of the interior stage image: the audience was seated in an open space on plain benches arranged around the playing area, which was filled with unmade beds, baby cots, blankets, open suitcases, potties, dolls and toys. Minimizing the distance between audience and actors by placing both on the same level and within a condensed space, Chilton intended to confront the usual escapist expectations of the Israeli audience: 'I do not want the audience to sit far away and think to itself – I am outside it, it does not concern me [...] theatre that people come to in order to run away from life does not interest me' (Naaman, 1981).

The scene opened with the actresses seated on a row of benches in front of the audience, each with bare feet and untidy hair, wearing casual clothing. One by one, the girls introduced themselves, giving a short autobiographical account focusing on a detailed description of their sex lives. At the end of each 'personal piece', the actress visually transformed herself into the specific character of a battered woman by means of props such as an Oriental kerchief, a rumpled housecoat or a broom. After recounting her dreadful life, the character removed the props that had symbolized her as a battered woman and returned to her persona as actress. This passage from 'life' to 'theatre' and back to 'life' was a poetic device intended to obscure the borderline between the real and the fictional and to reinforce the authenticity of the performance. Chilton also explained her choice in ethical terms: 'if the women are exposed to their bones then the actresses too should be exposed to a certain degree' (Naaman, 1981).

An invasive gaze

Chilton's genuine desire was thus to create a closeness and alliance between the Ashkenazi 'well educated, cultured and talented' (Joseph, 1981) acting graduates and the women from the shelter.

She tried to achieve this through the 'confessions' of the actresses at the beginning and closure of the performance, but the passage from the representation of woman as actress to that of woman as battered, and back again, instead operated as an apparatus that generated 'difference' and bi-polarity.

Chilton used fashionable images to signify woman as actress/artist, such as bare feet, jeans and unkempt hair, and then marked the transformative act through the adoption of common Mizrahi female images, such as an oriental styled kerchief, a frumpish housecoat, a broom or a 'baby' held in the arms. This created a conspicuous visual gap, which further extended the intellectual and emotional distance between the personal revelations of the actresses about their sex lives and the personal revelations of the battered characters about their sexual abuse. Against the background of the 'normal' sexual problems of the actresses (for example, introversion, childishness, the need for a steady relationship, the longing for a penis or the preference for casual sex), the stories of the battered women were striking in their sexual cruelty. The husbands were on the whole characterized as alcoholic, drug-addicted, criminal and sexually perverse: one had forced his wife to have sex during menstruation, another imitated hard porn movies and used strange and painful accessories, one had abused his pregnant wife until she delivered a baby with developmental disabilities, and another had abused his wife's daughter. The detailed graphic descriptions of the sadistic activities of the abusing Mizrahi husbands motivated the politics of the performance aesthetics, which portrayed the issue of the battered woman as a product of the eccentric, sick and criminal action of the Mizrahi man.

The poetic device by which the actresses were introduced as 'themselves' thus reinforced their distanced intellectual and social positions, and objectified the battered women as the focus of their invasive gaze. This process was underlined by additional aesthetic choices, which 'bestowed' upon the battered women a heavy Oriental accent, broken Hebrew and exaggerated gestures. They were also stigmatized as irrational and superstitious: 'Pity you didn't know me before. I could have taught you magic against this magic. [...] you take grass and stones, put them in the fire and dance and dance, you dance it on the person's head and it draws out all the spell.'[5] According to Dan Urian, this external representation was part of Chilton's comic strategy to produce 'liberating laughter' among an audience that was being confronted with a disturbing depiction of a harsh reality (Urian, 2000).

Despite Chilton's insistence that her presentation gave an accurate portrayal because 'this was simply the reality we found' (Naaman, 1981), I contend that comic relief was not in fact her conscious premise, but more a product of an unconscious stereotyped point of view, which led her to make certain artistic choices. The audience, who shared Chilton's conscious or unconscious patronizing attitude, decoded the representation of the battered women as humorous, often funny.

On a political level, the battered woman was displayed as an oppressed individual who was indifferent to the plight of another, similarly oppressed group: 'Arabs are not human beings [...] you can not educate the Arabs [...] they should be expelled immediately [...] this is a Jewish State.' These representational choices decentralized two important components of the narrative: one, that battering has never been an exclusively ethnic problem – 'Do you know how many lovers and doctors' wives are ringing and coming here?' – the other the social and cultural discrimination against the Mizrahi immigrants by the Ashkenazi veterans:

> Do you know what happened when we came to the country? We were put, at once, in a town named Shlomi. We arrive, and what do we see? Nothing. Darkness of Egypt and metal huts.[...] sometimes we had no water, so we used to run to the fields of the Kibbutz. [...] and my mother used to work for the Kibbutz in the cotton fields [...] and Passover came, and we were given special permission to have a hot shower in the Kibbutz. So we walked there one hour on foot. We got there, but no shower there [...] they were afraid we would mess their shower. [...] Once, abroad, a family was a family, a father was a father, a mother was a mother. I remember when we first went to school. We were nice, happy and tidy. The first thing the teacher said to us is 'here you can not speak Arabic! Here is not Iraq, not Tunisia, not Morocco, here you will speak only Hebrew.' They made us feel that everything we possessed should be erased.

The disproportion between the abundance of voyeuristic sexual reports and the paucity of other life materials not only created a distorted image of the battered woman but also substantially reduced

Chilton's 'super objective' for the performance – to raise audience awareness of the Mizrahi ethnic problem.

The documentary technique, which aimed at minimizing the gap between text and context, nevertheless emphasized the real social and cultural position of the speaking subject (actresses and director) in relation to the battered women. She, the spoken, made by theatre into an object, was in fact subjected not so much to an identifying gaze as to a humanitarian, merciful gaze in search of sensational moments that would shake the socially dormant audience.

Battered woman as speaking subject

Since the 1990s the expression 'battered woman' has entered the vocabulary of the mass media as well as that of academic and social welfare circles. Various measures have been taken to address the problem, such as the foundation of additional shelters, guidance and treatment services, better arrangements in hospital emergency rooms, new orders to the police, and new legislation against violence that has enabled the police and the courts to remove a violent husband from the home for a period of up to six months (Svirski, 1978). Nevertheless, as Barbara Svirski indicates, although the issue is now more visible in the media, it is still depicted as either a crime of passion or a family tragedy. 'It seems', she protests, 'that the problem has been conceived, over time, as belonging to the welfare services, and as such it no longer disturbs the existing order' (Svirski, 1978, p. 240).

Within this socio-cultural context *A Plague not Written in the Bible* performed by clients of the Centre for Prevention and Treatment of Domestic Violence in Herzlia, is a unique symbolic act through which battered women challenge the social order, and by making do with theatre suggest an extended feminist perception of domestic violence. The appearance of battered women as speaking subjects on the public stage constitutes a cultural documentation of a problem significant both to society at large and women in particular.

The creative process

Centre therapists who were acquainted with the work of theatre facilitator and director, Hannah Vazana-Greenvald, and the community playwright, Ora Habib, who had collaborated on an earlier performance about battered women, *Every Seventh Woman*, in Shkunat Hatikva, invited Vazana-Greenvald to the centre to facilitate theatre. The premise underlying the project was that community-based theatre, as a means of articulation that generated drama from within and to a given community, might serve as a new, advanced model for group therapy.

The centre takes a feminist approach to the problem of domestic violence, providing various individual and group treatments that together construct a therapeutic process. In the initial stage, according to the centre manager, Danielle Mack, the battered woman works to reach the point at which she is able to admit, first to herself and then to her peers, that she is a victim of violence. In the second stage, she acts to recover the mental strength to take responsibility for her own life (Mack, 2001). The frame of community-based theatre generates an additional circuit through which to expose a secret. This symbolic and public admission in front of the community marks the zenith of the therapeutic process. The clients as artists/actresses exhibit a sense of self-esteem and involvement, an ability to express a social statement, and a strong will to reintegrate into the community at large (Tagrin and Vazana, 1999).

A Plague not Written in the Bible exemplifies the creative process by which a therapeutic group of battered women made do with theatre in order to tell themselves by themselves. For almost a year, Vazana-Greenvald guided the women to articulate repressed life materials through dramatic exercises such as visual images, body movement, dough sculpting, storytelling, personal letters, monologues and role-playing. In the course of this activity the participants were invited to the Shkunat Hatikva Community Theatre to meet the actresses who had performed *Every Seventh Woman*. Later, the actresses from the centre hosted the group from Shkunat Hatikva. These encounters, which operated as consciousness-raising events for both groups, helped organize the centre actresses into a cohesive socio-artistic co-community able to realize their uniqueness as the first group of battered women to break the conspiracy of silence, literally as well as symbolically.

Vazana-Greenvald and Yael Tagrin, the social worker who accompanied the sessions, documented all the self-texts of the women, which later became the source material for the play written by Ora Habib, and which she later rewrote in response to the group's critical comments.

Performance analysis

A Plague not Written in the Bible depicts the daily experiences of women living in violent relationships. The narrative focuses on a liberating encounter between three sisters and their mother on the eve of Passover at the mother's home. In the analysis below, I expose the feminist materials articulated by the battered women that contributed to the discourse about gender and the dynamic construction of female identity, materials that symbolically deconstructed the prevailing approaches to the problem of battered women.

Community-based theatre in Israel is generally issue-oriented and tends to be realistic in style. *A Plague not Written in the Bible* appears to follow this pattern, and so it is in the realistic key that I initially decode the performance. The set reveals the interior of a very simple apartment, indicating the low income and traditional origins of the family. The kitchen, stage right, is represented by a few props – a low shelf piled with plates and a small round table on which are bowls filled with various fruits and vegetables. A green armchair and two smaller chairs are centre stage, and at the back there is a window with curtains, above which hangs a picture of a rabbi and an open holy script. At stage left is a simple iron bed with a figure lying on it, silent, totally covered by a blanket. The widowed mother and her daughters are busy preparing the holiday meal to be served that evening. Danielle, the oldest daughter, is a big woman, about fifty, married with children, attired in a long dress and wearing a hat, which indicates her religious lifestyle. In contrast, Josepha, the middle and single daughter, who has just arrived from Paris, is wearing a fashionable and sexy outfit, which suggests her mental and geographical distance from the family home.

At this particular family reunion, these women, encouraged by Josepha, manage for the first time to break the barrier of silence and to discuss with one another the sources of their distress. The mother finally admits that her abusive behaviour toward Josepha was the painful projection of a battered wife who could not produce boys. Josepha confesses that the violent relationship between her parents led her to behave like a boy, to run as far away from home as possible and to choose to be totally independent and single. Danielle reveals, for the first time, that her husband has been abusing her mentally for a long time. He avoids her intimate approaches, excludes her from the Sabbath service, harasses her when she talks on the phone, and has blocked their bank accounts so that she has to beg him for money. The story of the family, which is revealed step by step through dialogue between the three women, and through a few short flashbacks, creates a contrast to the behaviour of Igaella, the youngest daughter, who stubbornly continues to sleep, deliberately cutting herself off from her surroundings. The exposure of their mutual pain constitutes solidarity between the other three women, which empowers them to shake Igaella out of bed.

The final scene clarifies the structure and purpose of the whole performance, which has moved steadily from a conspiracy of silence toward the denouement of its total dismantling. Igaella, the pride of her family, a beautiful, educated woman married to a successful, well-educated Ashkenazi man, is a battered woman too. She acts like a typical abused woman, despite her education and professional achievements. She blames herself, trying to make peace with her husband and keeping her situation a secret. He has struck her even though she is pregnant. Bleeding, she has fled to her mother and is now hiding in the shelter of her bed. It is only in the last scene, encouraged by her sisters and mother, that she dares to rise and reveal her bleeding body and soul. On Passover eve, the traditional holiday celebrating the liberation of the people of Israel from Pharaoh, Igaella decides to take responsibility for her own life. The end of the performance signals the beginning of this liberating journey, which is supposed to free Igaella and her potential peers among the audience.

This realistic reading of the performance characterizes community-based theatre in general. But as this is community-based theatre produced by a unique group of women, I will also suggest what Dolan calls a 'resistant reading', which deliberately transgresses the explicit meanings of the text and in the process identifies the subversive tactics used by the actresses/director to deliver additional, often subversive, layers of textual meaning (Dolan, 1988).

Deconstructing myths

The choice of Passover eve as the temporal location of the narrative of *A Plague not Written in the Bible* is an obvious symbolic device anchored in the central liberation myth of Jewish-Israeli culture.

But in order to create a genuine parallel between the mythic liberation of the festival and the concrete liberation of the female characters in the play, the actresses, as poetic poachers, intervene and
disturb the mythos by making present its feminine absent part. The
title of the performance indicates that from the actresses' point of
view the act of battering women is as dreadful as the Ten Plagues
that God brought down upon Egypt. But, while the Ten Plagues
won figurative recognition in the Bible and have been put to various
metaphoric uses ever since, the symbolic silencing of battering is a
constant phenomenon that is only temporarily halted by the performance, which makes present the 'unspoken' plague that is absent
from the cultural discourse.

The first feminine image to appear on stage is that of 'woman
cleaning'. Danielle vigorously scrubs the table and polishes the silver.
This repetitive activity, which characterizes Danielle throughout the
whole performance, is consolidated into a visual motif followed by
repetitive sentences such as 'There is a lot of work now', 'There are
so many more things to do,' 'I haven't managed to do anything.'[6]
The actress obsessively polishes and re-polishes the cutlery, until
the realistic style of acting is disturbed by a Brechtian *gestus*, which
moves the spectator's attention from the textual signified to the sig-
nifier. To ensure that the audience does indeed perceive the image,
the mother delivers the interpretative function saying: 'What is the
matter with you? Cleaning like a crazy woman, enough!' 'Are you
crazy? Cleaning the house like a polishing machine.' In this way the
actresses create the subversive dimension of the accepted image of
'woman cleaning', which signals that from their point of view the
immense work expected from them on the eve of Passover is actually
the opposite of liberation, as the mother adds: 'Fifty years I'm preparing Passover and I don't rest for a moment. To rest now? At Passover?
Who rests at Passover?!'

Generally, the women prepare the holiday for the men, who
conduct the service without any symbolic reference to them. In the
Passover service that the battered women present on stage, they
articulate feminine experiences that are excluded from the official
traditional text of the holiday. Josepha says, for example, 'What a
smell! The smell of Passover, food together with detergents. The food
disappears after a day or two, but the detergents drug me for a week.'
When she realizes that Danielle has decided not to return home,

and that Igaella persists in sleeping, Josepha suggests two alternative options for the usual service: 'We'll sing songs, invite Chippendales who will dance for us, mother will cry for Dad, it will be fun. Maybe we'll join Igaella and we'll do the holiday in bed?' Josepha also replaces the content and order of the symbolic questions tradition- ally asked by the youngest child at the dinner: 'I have the first ques- tion honoring the holiday. Are you free [to her mother]? And you [to Danielle] are you free? Do you live in a great light, or in the dark- ness?' Josepha initiates a journey of revival, which indeed constitutes an alternative service, conducted by the mother who uses the tradi- tional service model to deliver a concrete liberation story of her real family in place of the mythic narrative: 'How is this night different from all other nights? You [to Igaella] will protect your baby, you will get out of bed, you will eat, you will say "I'm not guilty" [...] In what way is this night different? This night you will make a move, you will take responsibility for your life.' This alternative Passover service on stage is both a feminist challenge to the traditional male order and a public declaration of a new order of life by which the women reject all expressions of violence and move toward self-liberation.

Another myth challenged in *A Plague not Written in the Bible* has to do with the traditional perception of family. Throughout their lives women passively assimilate various axioms about how they should function within their families. Some of these indirectly support vio- lence against women, and lead them to a submissive acceptance of it: 'The unity of the family is important at any cost', 'The responsibility for the man's behavior depends on the woman' or 'Everyone gets what he deserves.' In the performance the actresses deconstruct the mythos of the family through the conservative character of the mother, who is the bearer of the popular ideology of the woman of valour. When Danielle tries to convince her mother that she has reached her limit in attempting to hold on to her abusive marriage, the mother blocks her: 'What kind of a mother leaves her kids? It's not like you. Do you want to cause harm?' She also attacks Josepha, her single daughter: 'A woman without a man? Lonely? How can a woman feel good with- out a man? [...] and if she becomes free, wouldn't she feel imprisoned inside herself? [...] Why should I open up my secret? Who will close it back? It will only leave pain.' The mother consistently refuses to contain her daughters' pain, instead reiterating the oppressive myths that had made her endure her own husband's violence and transfer it to her daughters. The turning point arrives when she confronts the cracks in the mythos of the non-violent, educated, beautiful bourgeois couple. When the mother realizes that Igaella's 'wonderful, excellent husband, I wish you [Josepha] such a husband' has beaten her daughter until she bled, she immediately abandons her patriarchal conformity and declares her determination to begin on an alternative, liberating path: 'It must stop now! Now! All of us were punished enough for things we didn't do.'

Miming the image of the femme fatale

Josepha, who has just arrived from Paris, bearing gift-bags full of silk and muslin lingerie, is wearing a tight red dress, high-heeled shoes and a red beret. She moves her body gracefully, with extrovert sexuality and expressive nonchalance. This female image is constructed at first glance in accordance with the basic patriarchal principal, displaying 'woman' as an exhibitionistic performance subjected to the male gaze. As such, Josepha is characterized through the basic male representational mechanisms: she behaves outrageously, displaying a daring style of dress and her fetishist accessories – the red high heels, red beret and glossy shopping bags. She is a femme fatale who must be re-educated and converted.[7] The part of re-educator is conventionally fulfilled by the mother, who tries to persuade her daughter to become a meek wife and good mother.

However, the construction of Josepha as femme fatale is also a form of cultural citation which, as Judith Butler indicates, is never an accurate repetition of the original image (Butler, 1990). Josepha as a critical character generates the narrative that exposes the citation not as a mimetic reproduction of the accepted femme fatale image, but as a mode of miming that ironically disturbs the image and subversively plays with it (Diamond, 1989). Josepha's ironic play inverts the inferior image of the femme fatale into a form of declaration that will sabotage the familiar image. She refuses to co-operate with her mother's regime of silence and denial, delivering oppositional statements with humour and self-awareness. She creates short shock effects, which not only break the conspiracy of silence, but also posit an alternative feminine lifestyle. She consciously enjoys her looks, exhibits a logical and determined thinking competence, a sharp tongue and leadership skills. She chooses to live independently, sometimes alone and at other times with a man whom she favours according to her own standards. She is open-minded, direct and adventurous, taking life in a playful and pleasurable way. The actress as Josepha actualizes a latent fantasy: 'I'm your dream, day and night you are praying to become like me.' By embodying this wishful thinking on stage, she validates it as a potential option, as a significant possibility within a new, more complex female identity.

Gender, according to Butler, is not a given, static socio-cultural construction, but a dynamic performative category, constantly generated in the course of action (Butler, 1990). While Butler refers to the performative stratum of everyday life, I suggest that in the symbolic context of the theatrical event the conceptualization of Josepha gains extra relevance. Her character is not a faithful reproduction of an accepted cultural image; rather the symbolic actions of the actress as character constitute a grounded aesthetics that generates a new image from within and in confrontation with the familiar image. Thus, the woman-in-red is a clear outcome of the way the actresses make do with theatre as an arena in which to contest the significance of 'gender'.

The difference in representation between *Battered Women* and *A Plague not Written in the Bible* is in part a result of the shift in public discourse in relation to the issue of battered women. Whereas at the beginning of the 1980s it was perceived primarily as a problem restricted to the 'uncivilized' Mizrahi co-culture, since the 1990s battering has became defined as domestic violence and has been treated as a more extensive and complex social problem. More significantly, however, the distinction between the two performances is an outgrowth of the identity and social positioning of the performers themselves. The shift from a distant, humanitarian speaking subject that stands for the battered woman, to the battered woman who speaks herself on stage, signifies the passage from acting woman as the object of a voyeuristic, invasive gaze, to acting woman as the subject who appropriates the gaze through the medium of theatre, and looks back, deep into society, with her own critical gaze.

The performance of *A Plague not Written in the Bible,* which was based on the actresses' own experiences, publicly recreated their transformative process, moving from the phase of self-accusation and denial to the phase of self-awareness, and then to the active phase of revealing the secret and taking responsibility for their lives.

However, this performance was not only the outcome of the total therapeutic empowering process that these women had experienced in the past through the various therapy sessions and the creative and production process of *A Plague not Written in the Bible*. The public performance also constituted an additional empowering process, 'here and now', that pushed the boundaries of the therapeutic into the socio-political arena. Performing their own resistant theatre in front of their family members, other battered women, and representatives from the welfare services and the academy, thus reinforced the actresses' determination to confront violence and marked their new, more extensive and positive social identity as feminist activists. Making do with theatre became the activist symbolic tool by which they strove to contribute to the discourse about gender in general and to the meaning of battering in particular.

7
Between Home and Homeland: Ethiopian Youth Making Do with Theatre in a Boarding School

In this chapter I focus on a group of Jewish Ethiopian youth at the Alon boarding school who were offered a theatre project as one of their recreational activities. In this case study my intention is to consider the politics of the boarding school as one of the settings of community-based theatre in Israel, and to investigate the extent to which a specific group of young black immigrants managed to make do with theatre for their own needs in such a location. The chapter opens with an examination of the general and the local social contexts and presents the construction of the Jewish Ethiopians as a new co-culture in Israel. It then considers the state's boarding school policy in general and the Alon boarding school policy in particular, setting this discussion against other nations' models of boarding schools for their native populations and the recent Australian government's declaration of regret over their past boarding school policy.

The second half of the chapter focuses on the performance text, from the creative process, including the scripting of the play, to the public performance and its reception. The discussion is based here on the discourse of Diaspora, which addresses issues such as ethnicity/race, nationhood, identity, displacement and belonging, all of which are of major concern here and contribute to understanding the complex 'diasporic condition' of the Ethiopian youth group in the boarding school as 'inside-outsiders' (Georgiou, 2006, p. 3).

Jews from Ethiopia as a new co-culture

The Israeli national meta-narrative reconstructs the history of the Jewish Diaspora upon the principle of 'from Calamity [Shoah] to Revival [Tkumah]' (Zerubavel, 1995; Shenhav, 2002). Linear time is cross-cut with Jewish experiences in different places, especially in Europe, North Africa and the Middle East, where the diasporic Jewish communities were repeatedly abused by the locals until finally, Israel, the saviour state, was established. Israel, according to the official narrative, is in fact the real, original homeland, gathering and reuniting the Jewish people who had been brutally scattered throughout the Diaspora. This mythic pattern was redeployed for the inclusion of the Ethiopian immigration into the national collective memory and master narrative.

Jews from Ethiopia arrived in Israel in two waves of immigration, in 1984 and 1991. The national narrative emphasizes the heroic saviour role of brave Israeli soldiers and airmen. Accordingly, the Ethiopian Jews were brought to Israel as the result of two dramatic, secret state operations. Israeli planes rescued the Jewish refugees from a harsh life and flew them safely to Israel, the homeland for which they had been praying, day and night, for over 2000 years. This miraculous modern exodus also alludes to the mythic exodus of the Hebrew people from Egypt, which the Jews cyclically retell every year during the festival of Passover.

The Jewish Ethiopians themselves underscore their own heroic part in this mythic exodus, telling of the long and arduous journey they began when they left their villages and townships and wandered for months during which they were attacked by robbers and suffered disease and starvation. When they eventually reached their temporary destination in Sudan, the severe living conditions in the camps there caused them additional physical and mental anguish. Finally, they reached 'the Promised Land', and, as one immigrant said: 'It was like after the Flood. Our ancestors had dreamt about the Holy Land for generations and here we managed to fulfil this enormous dream' (Pelled, 2007). Accordingly, the Ethiopian immigrants expected that the landing in Ben Gurion Airport, after such a long and terrible journey, would be the turning point for a new and better life.

The rhetoric of the Israeli national master narrative is, however, never quite equal to reality. The 'myth of arrival' encountered a difficult actuality in the form of the absorption process that awaited the Ethiopian Jews. This small community of 75,000 people disturbed the fragile status quo, and the Ethiopian refugees had to undertake yet another complex journey in order to break through the solid social barriers existing in Israeli society.

It is important to remember that Israel was established on the basis of a series of waves of Jewish immigration, especially from the 1950s. One of the largest waves was that of the Mizrahim – Jewish immigrants from Muslim countries, particularly from Morocco, Algeria, Tunisia and Iraq. These darker-skinned Jews from a totally different culture to that of the ruling Ashkenazi elite, were to be transformed by the policy of the 'Melting Pot', which meant that in order to become an accepted part of Israeli society they had to leave behind their own cultural heritage and identity and adopt the modern Western lifestyle that represented the new state. The second and third generations of these immigrants are still marginalized and fighting for full acceptance and representation in Israeli society. Although Israel, like other Western societies, could not fail to have become aware of the problematic implications of 'Melting Pot' policy, it nevertheless continued to make the same old mistakes, leading to physical, social and cultural segregation for the Ethiopian Jews. Moreover, in contrast to the earlier Mizrahi immigrants, who were immediately perceived as Jews beyond doubt, the Ethiopian immigrants have been forced to comply with Orthodox religious laws in order to become fully accepted as Jews. This cruel procedure was justified by the explanation that, in contrast to other Jewish communities in the Diaspora, the Ethiopian Jews had been isolated from the Jewish world for about 2000 years during which they had evolved and practised their own traditional form of Judaism. Thus, upon their arrival in Israel the official Orthodox establishment insisted on replacing their ancient religious tradition with the Orthodox accepted version (Ojanuga, 1993; Ben David and Ben Ari, 1997).

According to Baruch Kimmerling, a leading Israeli sociologist, this new immigrant group, as a new social category with no capital and mostly lacking modern skills, contributed not only to an even more diverse ethnic society than had existed before but, moreover, as a black group, introduced into the Israeli 'Euro-American' system an additional social phenomenon – a racial boundary that is almost impassable. Whereas in Ethiopia this community was perceived as relatively light-skinned, in Israel they have been marked as 'blacks'. In the local 'hierarchy of colour' the Ethiopian Jews are at the bottom, while above them are the Jews originating from Muslim countries and the Israeli Arabs. At the top are the Jewish immigrants of Euro-Ashkenazi origin and their descendants (Kimmerling, 2004).

The Israel Bank Report (2007) states that the Ethiopian community currently comprises 106,000 individuals, of whom 33,000 are native-born Israeli. Two-thirds of this community is made up of impoverished families with 5–6 children, a low level of education and still without professional skills. There are many single mothers, unemployed adults, high rates of crime, murder within families and suicide among the young. This process of social deterioration is explained from the Jewish Ethiopian's point of view as the inevitable outcome of the harsh absorption process:

> While at the beginning we perceived the gaze of the white veterans as an affectionate, caring gaze, we soon realized that behind those smiles there are people who think that we are nothing, that we are blacks, that we are not Jews. We were therefore told to abandon our culture, our values, and our customs. The Establishment behaves as if it has done us a favor. They would have painted us white if they could.
>
> (Pelled, 2007)

The young Jewish Ethiopians as youth-at-risk

Given these tumultuous socio-cultural and economic circumstances it is little wonder that most Ethiopian youngsters are considered to be youth-at-risk. A marginalized group tends to live in an ongoing state of threatened identity; thus, its younger generation, dwelling in such an atmosphere, is likely to develop an identity crisis (Shabtay, 2001).

It is a truism to observe that adolescence, with its accompanying biological, intellectual, emotional and social changes, is generally a difficult phase. Its primary – and always complicated – task is to establish the ego identity: 'a self image that organizes the past, enlightens the present, and directs future behaviors' (Chinman and Linney, 1998, p. 396). For young immigrants or young people born and raised in diasporic contexts, entering into adulthood is even more problematic. 'How will their memories of the homeland, marked by ambivalence and contradiction, operate? How will they relate to the cultural heritage of their parents? Will they reject aspects of the home country culture? Will they embrace other aspects?' ask Jana Evans Braziel and Anita Mannur in their introduction to *Theorizing Diaspora* (2003), highlighting the cardinal questions haunting Ethiopian youth in Israel. Adolescence for them is a concrete daily struggle since they live in a state of limbo. Obliged to acquire a new language and culture that excludes and discriminates against them, they are distanced from their families and community, whose culture is considered of little value in the dominant Israeli discourse, and who also suffer from deep poverty. The young Ethiopian writer, Omri Tgmalak Abarah, clearly expresses this, stating that he can neither identify with his parents nor with the Israelis (Pelled, 2007). In this liminal state the process of identity formation forces Ethiopian youth to the margins, engendering a sense of alienation and a deep identity crisis.

Malka Shabtay, an Israeli anthropologist who has studied the Ethiopian youth-at-risk, indicates how reggae and rap – black music – provides these youngsters with role models and with a means of retreat and survival. For these adolescents reggae and rap are more than simple musical preferences because they enable the creation of an autonomous sub-culture based around an identification with Afro-Jamaican and American music. This sub-culture also declares a desire to construct a new Afro-Israeli identity that separates them from both their community's tradition and the dominant Israeli culture and society (Shabtay, 2001). The intergenerational and intercultural gaps provoked by this arouse strong emotions – grief, anger, shame and frustration – that push the youngsters into negative modes of behaviour. 'The explosion will soon arrive', warns Baruch Dego, a leading young Ethiopian football player, 'too many Ethiopian youngsters fall into crime, drugs and suicide. Something has to be done' (Kobowitz, 2007). One of the central policies of the Israeli establishment towards dealing with the problem of immigrant youth-at-risk is that of offering them the option of boarding school as an optimal solution.

The Alon boarding school as the setting of the theatre project

From its inception the State of Israel has perceived boarding school education as the best way to absorb immigrant youth. Until the end of the 1960s most of the pupils in such boarding schools were young Mizrahi immigrants. Later on, the boarding school became an option not only for young immigrants but also for youngsters from poor families, again mostly of Mizrahi origins. The institutional view was that the boarding school provided a powerful setting for the disengagement of the youngsters from their families and their traditional cultures, while at the same time exposing them to the 'appropriate' cultural climate of the state (Kashti and Arieli, 1997).

The Israeli state boarding school policy recalls coercive attempts by former Australian, Canadian and US governments to assimilate indigenous children into the Western socio-cultural order. In the late 1800s the US government opened special boarding schools for the assimilation of young Native Americans, with the aim of training them for domestic service and farm work (Adams, 1995). Almost at the same time, Australia and Canada undertook similar but more vicious policies. Between 1910 and 1970, Australian Aboriginal children, mostly under five years old, were taken forcibly from their parents by police or welfare officers to special boarding schools where the declared intention was to educate them in order that they should be enabled to assimilate into Australian society. In the event, they received only a low level of education and were frequently subject to physical, mental and sexual abuse. As in the American model, they were in fact expected only to go into low-grade domestic and farming work. From the 1870s to the 1970s, the Canadian authorities forcibly relocated about 150,000 native Indian children into distant boarding schools, with the intention of detaching them from their native language and culture. More recently there has been a tendency in some Western societies, especially in Australia and Canada, to publicly regret this policy as a violation of human rights and as an oppressive, racially discriminatory apparatus.[1] The Israeli policy of assimilation through institutional boarding schools has always been a softer version of these earlier examples, and since the 1970s the patronizing educational model has become more open and empathetic in relation to the source culture of the children, especially in relation to the comparatively recent immigration of the Ethiopian Jews (Kashti and Arieli, 1997). Nevertheless, the state still uses boarding schools as an assimilation device for Ethiopian youth, concealing its intention by declaring that such schools offer more opportunities for these children than their parents can provide.

In any case, it seems that the state ignores the oppressive characteristics of the boarding school as 'a total institution' which separates the children from their home and community and displaces them into a closed and self-contained universe based on clear power relations and national cultural domination.

Ethiopian parents broadly welcome the boarding school option for its financial support and as a beneficial educational framework for their children. The candidates and their parents are never forced to accept the offered boarding school place. Indeed, some families (although not many) who have managed to come to terms with their low status prefer to keep their children with them and send them to good schools nearby.

The Alon boarding school is located in a kibbutz in the centre of Israel. It has been operating for 11 years and houses around 40 Ethiopian teenagers aged 13–18, who are taught in the local community high school together with about 800 pupils from the veteran families of the area. During the day these Ethiopian youngsters are integrated with other young Israelis, while in the afternoons and evenings they return to the boarding school where they are assisted by a social worker, a psychologist and several instructors, who help them with their homework and also provide them with various recreational activities, including football, drumming, singing and horse-riding.

The pedagogic policy of this institution, which seeks to sound more pluralistic and culturally open-minded than the old boarding schools, reflects nevertheless the hegemonic aims of the state, encouraging the Ethiopian pupils to achieve excellent results in their studies, then to enlist in the army, and eventually to become leaders in their own communities. From the point of view of the boarding school management, such a pattern of achievement would constitute a clear sign of the successful assimilation of these Ethiopian youngsters. Accordingly, the boarding school's internal social structure is actually separative in nature. The Ethiopian students join the regional high school in the morning, dispersed among the various classes according to age, while for most of the rest of the day and the night they live alone, as an isolated group in – as it were – a comfortable cage. In the boarding school the state of limbo experienced by the young Ethiopian immigrants is reinforced, since they live in a mental, emotional and physical vacuum. These were the specific local circumstances in which I sought to study the potential of community-based theatre to articulate and confront this vacuum, while taking into consideration that this cultural practice operates between the ongoing tension between top-down institutional integrative aims and bottom-up needs and desires, as well as reflecting this tension.

The theatre project was in fact spontaneously initiated. The principal of the high school, who is also the manager of the boarding school, approached the theatre teacher at the high school one day and asked her if she would agree to guide a theatre group in the boarding school. 'I don't think that this idea stemmed from some "visionary" belief in relation to the Ethiopian youngsters', recalls the theatre teacher, Chen Elia. 'She [the principal] was not familiar then with community-based theatre. Probably she had some extra budget and thought of an additional recreational activity. "They need to be more occupied", she said to me' (Elia, 2007). While the principal's proposal for a theatre project had arisen from a sense of good will and empathy, albeit still underlain by an essentially patronizing standpoint, Elia, was well aware of the subversive political potential of community-based theatre.[2] Consequently she accepted the offer and decided to employ this model of theatre with these youngsters. She managed quite quickly to recruit 12 youngsters aged 15–18. Two of them dropped out during the first stage of the creative process, while the others (5 boys and 5 girls) eventually put together a performance entitled *Frames*.

The creative process

Elia indicated that from the beginning of the project she was very impressed by the young Ethiopians' artistic and mental skills:

> My first experience with them, as a native-born Israeli who was not familiar at the time with the Ethiopian community, was that they are very talented, that they have natural theatre skills. They immediately understood my instructions. When I asked them, for example, to work with a prop such as hats and change their personality each time they put on a different hat, I was impressed by their smooth transformation into various characters. They were also very musical, and when I asked them to create something different from a familiar song they were enthusiastic, and introduced

different rhythms and lyrics that presented in depth their dreams and problems. From the beginning they showed their ability to be honest and frank. I was surprised by the openness of their minds and body movements.

(Elia, 2007)

In one of the earliest encounters each of the participants brought along his/her favourite song on a disk and danced spontaneously, leading the entire group to exactly copy the moves. The whole group then sat in a circle and collaboratively improvised, introducing different lyrics in different rhythms. This exercise later became the basis for the rap song in the public performance. In another encounter each participant was guided to create one scene based on a favourite song. The scene had to present a specific location, with characters as well as the song itself as an integral part of the scene. This exercise developed into several sessions in which the participants were introduced to the practice of theatre; they were trained to build up a character by using a specific object and/or by foregrounding a specific bodily gestus, and also practised various forms of improvisation. The scenes created by the participants functioned to elicit various fragments of previously repressed life materials. In some instances the participants returned to Amharic – the traditional language of their parents. At times they spoke Hebrew in a rap style. The scenes brought back to life traditional songs and legends from Ethiopia that the youngsters thought they had forgotten, as well as elements from the daily lives of their families in Israel. The facilitator suggested words such as 'home', 'father', 'mother' or 'boarding school' as entry points for encouraging the young participants to articulate their own life experiences. For Elia, as a native Israeli, these were well-defined categories, while for them, as 'diasporic subjects' (Braziel and Mannur, 2003, p. 5), these concepts were still 'in-the-making' and full of ambivalence and contradiction (Lavie and Swedenburg, 1996, p. 16). Myria Georgiou points out that 'home' is the domestic and familial 'starting point for identity construction and socialization'. For diasporic populations 'home has a more complex meaning than in other populations' (Georgiou, 2006, p. 6). In the case of the young Ethiopian participants in the theatre project the question of what/where is home is still at stake and it is this lack of a solid starting point for identity formation that provides an explanation for their identity crisis.

At this stage the scenes indeed reflected the young Ethiopians' frustration and disappointment in their parents, and in particular in their teachers and instructors at the boarding school. They felt that the staff were patronizing, indifferent and alienated.[3] This first stage of the creative process functioned as an incubation phase, providing them with a closed, intimate and contained environment in which to purge their muted feelings and thoughts in a way that no other recreational activity had done.

The second stage of the creative process was dedicated to the consolidation of the materials into a script and its production as a performance. It became apparent that while the youngsters were willing to publicly present materials in relation to their parents' traditions and daily life in Israel, they totally refused to expose their feelings in relation to the boarding school. They said, 'we shouldn't say these things, we better leave them out of the show, we don't feel we can do that right now' (Elia, 2007). The fact that these youngsters chose to keep silent about one of the basic issues (student-teacher relationships) of boarding school in their community-based theatre, reveals something of their fragile, insecure and detached state-of-being.

As part of the scripting process the facilitator gave each participant a sheet of paper with two opening sentences to complete: 'When I look at the mirror I say to myself...' and 'I would like people to think of me...' This exercise motivated the participants to write monologues which influenced and directed the script. One participant, for example, related to the suicide of one of the Ethiopian graduates of the boarding school: 'I can not forget you [a Hebrew curse], I can not forgive you. Wait, you son of a bitch, we'll meet some day in heaven. Why didn't you think of us, we were friends. I'm so angry. The instructors here try to hide everything, as if nothing has happened. Lately, I hate them – all of them.' Another wrote: 'I wish my friends in the neighbourhood would understand that I haven't abandoned them and come here because I have no interest in them. I wish they would understand that I can't become a part of their negative activities.'[4]

The scripting process and the rehearsal phase further united and bonded the participants into a cohesive group. This process was intensified at this stage by an additional socio-artistic activity initiated by the facilitator. The participants attended weekly meetings with a group of Ethiopian children at the cultural centre in a small

town near the boarding school. According to Elia's notes the participants found much pleasure in tutoring and facilitating theatre games
with their younger counterparts. This activity improved their theatre
skills and encouraged them to continue working on their own theatre performance.

As the scripting process reached its final stage, it became clear
that these youngsters were emphasizing the conflicts that they
had with their parents/family/community. Was this dilemma more
agonizing for them than the racial conflict with Israeli society? Elia
commented that when she began the project she had expected the
racial issue to arise quickly, but the work process revealed that the
Ethiopian youngsters at the Alon boarding school preferred first to
handle their intragroup problems and only then to confront their
intergroup difficulties. For me, as a scholar of the history of community-based theatre in Israel, this was quite a surprise. Back in
the 1970s, in the early years of community-based theatre in Israel,
the field was distinctively occupied by groups of radical Mizrahi
youngsters from disadvantaged urban neighbourhoods. They appropriated theatre in order to articulate their resistance to the process
of social stratification, which had been based from the beginning of
the Jewish State on the correlative principle between ethnicity and
class. This ethno-class process positioned the Mizrahim in the lower
socio-economic strata of Jewish society, but also produced a cultural
stratification. The young Mizrahi performers employed theatre as
a symbolic weapon with which to challenge their socio-cultural
and economic status as well as their negative identity, all of which,
from their point of view, was not of their own choice but had been
imposed upon them. Community-based performances such as *Joseph
Goes Down to Katamon* (1972) and *The Other Half* (1974) were clear
anti-establishment community-based theatre, protesting against
the Gordian knot that bound ethnicity and class in Israeli society.
Only later, from the 1990s on, as Israeli society progressed toward
more cultural openness and relativism, did the Mizrahi community-
based theatre groups become more reflexive and self-critical. In the
community performance *Phachme Dast* (1997), for example, the
performers openly acted out conflicts between themselves and their
parents, seeking to stimulate intragroup processes of transformation
as well.[5] With the young Ethiopian immigrants at the Alon boarding school, the youngsters seemed to be taking the opposite route.

Studying the public performance of *Frames*, its reception process, and its influence on the subsequent theatrical work, might explain this direction.

Frames: performance and reception analysis

The title of the performance, *Frames*, is in fact the interpretative key to decoding the performance in general and the central image on the stage in particular. The performers did not use actual frames but, nevertheless, in many instances in the performance, they evoked the frame of a picture, a mirror or a tableau through their physical gestures. This absent/present image accumulated into the super-metaphor of these youngsters' predicament. The complexity of their own lives lay not only in moving sharply in and out of binary frames, which youngsters may find difficult in general, but also and mostly in signifying and making sense of these frames, none of which was truly natural to them. Immigrating to Israel as infants or having been born in Israel means that Ethiopia, their country of origin is lost to them; at the same time, Israel, the mythical 'homeland' is still not and might never be 'theirs'. This disjointed experience was articulated through the episodic structure and style of the performance, which consisted of 10 short scenes and songs performed in a simple and straightforward manner. The stage was almost bare, with only a few props: a bed with a red cover and a table with a brown tablecloth pushed to the backdrop and a few white plastic chairs which were brought to front stage as needed. The performers were all dressed in black except when representing 'a parent', 'a neighbour or a traditional ritualistic element. In those cases they put on the *natala* – a white Ethiopian robe. The *natala* became the leitmotif in the performance, representing the youngsters' parents, community and country of origin. The episodic structure and style was characterized by a twofold division of colours (black and white). There were two kinds of scenes: pseudo-realistic 'here and now' scenes that articulated critical life moments of the Ethiopian community from the youngsters' viewpoint; and stylized 'then and there' scenes in which the performers not only represented but actually practised and studied elements of their original Ethiopian culture. In the latter scenes the performers used the Amharic language, traditional storytelling, music and dance and rap. They thus employed theatre in order to revive and remember that which they had never known or had since chosen to forget. Such a theatre work can operate as a potential starting point for constructing some kind of collective memory.

The first scene was staged as *tableaux vivants* in three 'friezes'. Initially, the performers stood motionless behind frames, dressed in the *natala* that represented traditional Ethiopian clothing. This tableau symbolized 'tradition', which equates with 'parents', 'family' and 'community'. The second frieze presented two frames: half of the performers, dressed in the *natala*, stood behind one frame, while the other half, dressed in ordinary dark clothing, posed as a humorous school photograph behind the other. In the third frieze the 'traditional frame' was quickly evacuated, leaving only two 'parents'; while the other, 'boarding school/society frame', became filled with cheerful young Ethiopians. Although there is no indication in the notebook that Chen was influenced by Augusto Boal, I nevertheless find it useful here to employ Augusto Boal's *image theatre* as a basic analytic tool with which to interpret this performative pattern. Boal suggests a transformative concept of image theatre in three steps. First the spect-actors[6] form a group of statues that imagistically display their collective standpoint around a central issue of their life. This image usually reveals the power relations that constitute them as a co-community. Boal defines this image as the *Real Image* (that is, the image of reality, the world as it is), which is always the representation of an oppression. In the second stage of the image theatre the spect-actors create the image of the desired world in which the conflicts have been resolved. This image is labelled by Boal the *Ideal Image* (the image of identity, the world as it could be), in which the oppression will have disappeared. The most important stage of the image theatre is the third and last one. It is here that the spect-actors take over responsibility and commit themselves to the process of modification 'in order to show in a visual form how it may be possible to move away from our actual reality and create the reality we desire; they must show the *Image of the Possible Transition*' (Boal, 1992, pp. 2–3; emphasis in original). Accordingly, the opening scene of *Frames* can be depicted as the first stage of Boal's real image, in which the young Ethiopian performers sharply posit the central conflict between themselves and their parents, between the binary oppositions of two homes, communities and identities. At the end of the performance the youngsters again appropriate the tactic of the image theatre, now presenting frozen frames full of smiling students, teachers and parents. The young Ethiopian performers thus create as closing images an idealized representation of the everyday life they desire. As such, it is interesting to look at the eight scenes between the real image and the ideal image as a particular mode of the transitional image, and to try to trace the process and characteristics of this modification.

The second scene was in realistic style and showed an Ashkenazi social worker visiting an Ethiopian family comprised of a father, a mother, a grandfather and a young daughter. Traditionally, Ethiopian Jews live in large family groups and, as a result of the allocation to them in Israel of small apartments, they often suffer from severe overcrowding. This was subversively signified on the stage by a humorous image in which three joined plastic chairs functioned as a 'sofa' on which all four members of the family were trying to sit. The encounter between the social worker and the parents was also presented in an amusing manner by using Amharic words that sound similar to Hebrew words but carry totally different meanings:

> The mother: Would you like to have *ingara* (a special Ethiopian bread)
> The social worker: No, thanks, I drank in *Aroma* (famous Israeli chain of coffee shops).
> The grandfather: Orumu? I knew one of them. Do you know these people?
> The social worker: Aroma, I meant Aroma. Do you know that your daughter was suspended from school? (the word suspended in Hebrew sounds very similar to the Amharic word that means to lie).
> The father: Did you come to tell me that my daughter is a liar?[7]

The tactic of grounding this element of the performance in comedy enabled the performers to articulate their ambivalent feelings and criticism toward the social worker who, as the representative of the powerful establishment, insensitively used the Hebrew name of the daughter, Tamar, instead of her Amharic name Bortoken, and is entirely convinced that the best place for her is in the boarding school. The father is not eager to send away his 'helping and cleaning' daughter. But eventually, after the social worker promises that the boarding school is a nice place that provides a good education, he agrees, and the daughter thanks him in Amharic. Here, the performers implicitly criticized the patriarchal power relationships in the Ethiopian family, while explicitly pointing to the universality of a generation gap in which parents worry whenever their daughter is out of their sight and might be tempted to 'drink alcohol, do drugs, do boys...'

By speaking in two languages, both literally and metaphorically, the young Ethiopian performers not only articulated their deep feelings of ambivalence and contradiction, but at the same time they used the stage to open up communication channels in order to bridge the divide between themselves and their mixed audience as well as between the Israelis and the Ethiopians in the audience.

The ambivalent position in relation to the parents was further articulated through the image of a phone conversation between Bortoken/Tamar and her mother. First, they stood close to one another, face to face, and spoke Amharic. Then they started to move away from one another, talking half Amharic and half Hebrew. By stage three they were standing distanced from one another, with the mother talking only in Amharic and the daughter only in Hebrew. Finally, the daughter exited, leaving the mother totally alone on stage. This detachment from the parents as 'home' was followed by a stylized episode celebrating the reattachment to Ethiopia as 'home'.

It opened with a song in Amharic called 'Ethiopia Agre', which expresses a deep longing for Ethiopia, the old homeland. Back in Ethiopia these youngsters' ancestors used to sing 'Jerusalem Agre', praying and longing for Israel. The performers had heard this song at home and put it on the stage in their own version, seeking to reunite through the singing with their lost identity origins. Here again, the question of what and where is home/homeland is raised. After the song each performer went down into the auditorium and up onto a smaller stage, each held a flashlight directed at their face. Each performer told his or her own story in traditional style to the spectators sitting nearby. The principle of such storytelling is that each story starts with the words 'tarat tarat', which means in Amharic, 'Once upon a time.' This form of storytelling had been a traditional ritual back in Ethiopia. Every day after sunset the community used to gather and the adults would recount their legends to the children. In the performance the youngsters used the old ritualistic frame to recall both these legends and the stories of their parents' passage to Israel. For example:

> Shlomo: Tarat, tarat, we immigrated to Israel in 1991. We came without my two older brothers. Why? I do not know exactly. Maybe because of the mess in the camp...They immigrated in 1995 but we didn't meet until 2003! They were looking for us all these years and we didn't know. We were not informed. We thought that they will never come. We were afraid that they might have died. One day, when I was playing football my mother called me to come up. She said she had a surprise for me. I saw a few strangers and couldn't figure out who they were. I was told that they are my brothers. This was the surprise, it was exciting to meet them.
>
> Masganao: Tarat, tarat, one student studied for 6 years but couldn't pass the exams. Three times he tried and three times he failed. The student was distressed and decided that he was not fit for school and had better look for a job. On his way it was raining and he sat under a tree waiting until the sky would be bright again. Suddenly, he saw a small beetle trying to mount the tree trunk. But the trunk was wet and slippery and the beetle was climbing and falling, climbing and falling...She tried six times and on the seventh time she succeeded! Then the young student said to himself: 'If such a tiny beetle does not get desperate and at the end could make it, I can succeed as well.' He returned to school and continued studying. Indeed, he not only succeeded but, after a few years, became the principal of the school.

This scene offers another clear example of how the young performers appropriated theatre not only in order to engage in some kind of memory-building but also to practise their chosen traditional elements in the here-and-now. Moreover, considering that the audience was comprised of these youngsters' family members, the teaching staff and other official representatives, this storytelling had two different functions and effects. For the family members, who knew the style as well as the stories, it functioned as a twofold bridge, lessening the sense of alienation that had arisen between the parents and their children. For the Israelis, who knew almost nothing about Ethiopian traditions, it functioned not only as a tool for cultural dissemination but also as a platform to publicly exhibit the complex identity formation of these youngsters.

   This somewhat nostalgic and optimistic scene was disrupted by rap singing, unfolding the harsh life story of Efraim, a young Ethiopian, who was in fact the prototype of their collective biography:

> Efraim was an innocent kid
> He arrived 7 years ago
> He lived in a distressed neighbourhood
> A very difficult place
>
> Although his parents didn't think
> It was a problem
> Efraim made up his mind
> And moved alone to the boarding school
>
> He managed alone
> He learnt about reality
> In the hard way
> He learnt to count on himself only
>
> Efraim loves life
> He remembers his home
> He thinks about his old friends
> He learnt to combine boarding school with parents

Is this indeed their wish? To combine boarding school with parents? The following realistic episode which showed Bortoken/Tamar, who has moved to the boarding school, visiting her home, reveals how difficult this combination is for a young Ethiopian who moves geographically, mentally and emotionally 'betwixt and between' their family and the boarding school. On the sofa this time two more women were crowded. They are two gossips who functioned like a Greek chorus, representing the community's oppressive standpoint:

> Gossip 1: Have you heard? They sent their daughter, Bortoken to the boarding school.
> Gossip 2: Boarding school? They are bad.
> Gossip 1: Yes, have you heard what happened to Efraim?

Gossip 2: Of course I heard. He is on hashish and drugs.
Gossip 1: And what about beer and cognac? In spite of this they sent their own daughter to the boarding school.
Gossip 2: I keep my daughter besides me. She cleans washes and cooks. That is a good girl!

This text, which was performed in a broadly humorous and exaggerated manner, articulates the critical gaze of the 'inside-outsider' Ethiopian girl who, as a boarding school inhabitant, acquires a 'dual vision' in relation to both her family/community and the boarding school. From Bortoken/Tamar's point of view the adult women would like to bind her to the traditional oppressed way of life for a female. Tamar, who arrives from 'there', dressed in jeans and a fashionable blouse, wearing sneakers and talking on her cellular phone, not just in Hebrew but in slang, is received as a stranger. Her community is looking for their familiar child, known to them as Bortoken, and can not cope with the new 'Tamar'. Her father cannot comprehend what it means to go on a field trip: 'We have sent you to study not to stroll', he says angrily. The scene ends with Tamar crying: 'You never understand me, you don't understand my world, and you know nothing.'

But to what kind of a 'world' does she refer? Her world is still in fact a fragmented and contested world. Thus, when she returns from 'there' to 'here', she is not yet sure where is 'there' and where is 'here'.

This inner conflict was articulated next by an image presenting the performers standing in a row, in front of the audience, as if in front of a mirror. They were all wearing the *natala*. Each of them then slowly takes it off, preparing to go out, when suddenly the rest of the group notice that one of them has chosen to keep his *natala* on. 'What happened to you? Don't be that heavy', shouts one. 'Hang loose! Don't you want to be like everybody else?' shouts the second. 'Change your clothes, you look like my grandfather', shouts the third. 'Don't you want to have style? You are not in fashion', shouts the fourth. 'Maybe I'll buy you some clothes; we are here in Israel, wake up!' shouts the fifth. All these sentences were said repeatedly, in an accelerating rhythm and voice. But, nevertheless, the performer with the *natala* blocked his ears and remained as he was! If the impression gained was that this determined youngster was signalling a small

positive sign of progress toward reconciliation with his parents' tradition, then the following scene showed that what had just been seen
externally was emotionally decisive and even predicted disaster.

The eighth episode was the most intricate, touching upon the
issue of suicide, which is high amongst Ethiopian youth. This stylized scene was built on a series of daily activities being carried out
simultaneously on stage, and focused on the moment when each of
the performers finds out that Haim, one of their friends, has committed suicide. One individual is taking a shower, a second is arranging
boxes, a third is playing ball, a fourth is watching television, a fifth
has just arrived back from a trip to Ethiopia, a sixth is reading a newspaper and so on. The newspaper and the television are full of items
reporting the terrible events in relation to the government's plan for
the Gaza Disengagement. Nevertheless, these youngsters' shock is
not the result of this national trauma:

> I heard a woman shouting on the television 'this is the darkest day
> in History of Israel,' and I shouted at her 'maybe for you but not
> for me. It has nothing to do with me.' Then my father and sister
> came in and told me about the suicide. Then it became my own
> darkest day. How could it happen? How could it be one of us? I
> couldn't believe it was one from the Alon group.

These youngsters employ theatre not only to commemorate Haim
but also to act out their deep sorrow and shock: 'I can't believe it,
leave me alone', 'I know you are at peace now. I wish I would have
been there for you.' 'I'm furious with you. It is not fair!' 'Why? Why?
I ask you why?' 'Pity I didn't phone you to say a word to you. You
were a good friend. I love you and won't forget you.' At the end of
the scene the performers symbolically reburied Haim. The performer
who in the previous scene had not removed his *natala*, now lay on
the stage floor, while all the other performers covered him with their
natalas, declaring: 'Haim Aragy, 17 years old, left behind 4 brothers
and a lot of dreams.'

The next scene functioned as comic relief from the previous episode, seeking to colour the whole performance with a more vivid and
optimistic tone. The episode depicted 'parents' day' at the boarding
school. This was a utopic manifestation since the Ethiopian parents
hardly ever come to visit the school. Moreover, the scene presented a

harmonious, humorous and communicative encounter between the teachers, the parents and the students:

> Mathematics teacher: I'm so happy to meet you. Tamar is very nice although she has some difficulty in mathematics.
> Father: Well, Tamar, that's no problem. Here is one sheep and here is another and that makes two of them. One we ate, hop she disappeared!
> Geography teacher: How wonderful to meet you. You should come and tell your immigration story. It has a lot of geography in it.

At the end of the episode, after each teacher has been fascinated by the simple and helpful suggestions that the Ethiopian father provides, the parents admit: 'Tamar is a good student, she knows a lot', and Tamar answers: 'There is so much more I can learn from you.' This moment of reconciliation on stage was followed by the idealized image in *tableaux vivants* of smiling parents together with the students and teachers. Did this indeed refer to utopia from the performers' point of view? It is important to emphasize that the transition between the real image at the beginning of the performance and the ideal image at its closure was not structured as a linear, progressive route. It was rather manifested as a pendulum, swinging forward and backwards between 'Ethiopia' and 'Israel', 'home' and 'homeland', or maybe vice versa, exposing the problematic struggle of these youngsters to retain some elementary codes for their everyday life. By successfully acting on stage white and black, young and old, feminine and masculine characters, the performers exposed the socially constructed quality of reality, and laid bare the deliberate inflexibility of the social status quo.

The public performance and its reception is usually the catalyst phase of the whole community-based theatre project. *Frames* generated a festive and cheerful atmosphere and, in general, a delighted response.

> The high-school teachers, the social workers and the boarding school instructors approached me with tears in their eyes. They said to me 'How did you manage to motivate them to open themselves, to speak up, to tell you their stories?' They were so excited, indicating that the performance had introduced them to

the Ethiopian youth in a new light. They said that now they had begun to understand things in relation to this group that they couldn't comprehend before. They were mostly thrilled by these youngsters' artistic talent and intelligence, and found it important to carry their impressions to the other veteran Israeli students of the high school. The principal of the high-school and the boarding school approved on the whole of the performance and decided to continue the theatre project by any means.

(Elia, 2007)

These reactions, clearly positive and important for the continuity of the theatre project, reveal, however, the unconscious deep-rooted racial perceptions of the boarding school staff.

According to Elia's notebook the Ethiopian parents and relatives of the performers were more reserved in their outward reactions. Some of them approached Elia and said mildly that they had enjoyed the performance. Others stood smiling, hugging their children and talking with them in Amharic. The performers were greatly excited, enjoying the warmth of the attention and the compliments that they received from all those who had attended the performance. They later told Elia that the performance had actually left their parents torn between laughter and tears, and that it had also confused them, as they had only partially understood the Hebrew. They had tended to react to the humorous elements of the performance and preferred to show indifference to the more problematic scenes. This, they explained, was part of their traditional behaviour of speaking only about the good things.

Discussing the universal difficulties of the developmental period of adolescence, Chinman and Linney (1998) propose a model of empowerment that can serve as a preventive intervention for many of the problems of adolescence. They indicate that active participation, awareness of the surrounding world and identification of strengths, which are key components of the empowerment process in general, are also developmentally important during adolescence (395). They suggest a model for the adolescent empowerment cycle in which 'adolescents are engaged in a process to develop a stable, positive identity by experimenting with different roles and incorporating the feedback of significant others' (398). Participating in positive, meaningful activities, learning useful and relevant skills, and being recognized, are the basic aspects of the empowerment cycle. Chinman and Linney (409–10) assert that:

> the process of empowerment, or the structure of the empower-ment cycle, will result in positive or negative behaviors depending on the kind of participatory experiences available to the adoles-cents. Thus, the nature of participatory opportunities and the specific roles in which the adolescent participates may define the valence of empowerment (positive v. negative), and the nature of the experiences will depend in large part, on the community context in which the adolescent lives.

Community-based theatre is a cultural practice that fits this proposed model and has the potential to generate its empowerment cycle. The theatre project, as demonstrated here, provided the Ethiopian adolescents with participatory, meaningful activities and roles through which they acquired the skills of theatre and team-work, and achieved recognition from both their family members and the educational staff.

These Jewish Ethiopian youngsters from the Alon boarding school are a unique and specific group of immigrant adolescents. However, the contemporary epoch of globalization is characterized by mass migration which constitutes new forms of Diaspora (Appadurai, 1996), and groups of immigrant youth-at-risk are a widespread phe-nomenon today in many disadvantaged suburbs of Western cities and as such share common behaviours and life-styles. The group at the Alon boarding school is part of this global phenomenon, but its state of being is even more acute since, according to the Israeli national-Zionist narrative, Jews from the Diaspora are not immi-grants but are, rather, returning to their ancient homeland in order to live in harmony with their brothers. Consequently, the harsh real-ities of absorption as described in the opening part of this chapter, face these Ethiopian immigrant adolescents with acute disappoint-ment, despair and disengagement. Their limbo state is worsened in the boarding school, which provides them with comfortable yet confined conditions. Here, the community-based theatre project, which was basically initiated as a recreational activity, succeeded in generating a more elementary, fundamental tool for everyday life and identity formation. If identity, as Stuart Hall indicates, is a 'production' which is never complete and always constituted within representation (Hall, 2003, p. 234), then community-based theatre is that representational practice that enabled these Ethiopian youngsters to produce their own process of identity formation.

Community-based theatre managed to present and problematize the ambivalent and displaced lived experiences of the young Ethiopian performers as diasporic subjects, and provided them with a critical, reflexive site from which to confront this existence. Diasporic experiences, as this theatre project revealed, are not essential or pure but heterogeneous and diverse. Thus 'homeland', 'home', 'nationhood', 'belonging' or 'identity' are not solid categories as the Israeli master narrative imagines, 'but processes always in change and always mediated by issues of class, ethnicity, gender, and sexuality' (Braziel and Mannur, 2003, p. 14). The theatre project of the Alon boarding school does not suggest that community-based theatre is some kind of quick, magic remedy. It is, nevertheless, a unique 'third space' that engenders creativity, identity formation and cultural negotiation; and always engenders hope for a better society.

8
Undoing Political Conflict: Israeli Jews and Palestinians Co-Creating a Theatrical Event

In 1998, the cultural centre of Ramle, a mixed town of Israeli Jews and Palestinian citizens in the centre of Israel, initiated a theatre project in collaboration with the Theatre Department of Tel Aviv University. This eventually evolved into a five-year project. In 2002, another project was undertaken with Jewish and Palestinian students in Jerusalem, initiated by the workshop on 'Theatre as a Mediation Tool' and co-sponsored by the Billy Crystal Project: Peace through the Performing Arts. These two projects exemplify the complexity of facilitating theatre with a mixed national co-community in a place of conflict such as Israel. Before discussing these projects I shall present their political context: namely, the position of Palestinians in Israeli society.

There are currently more than a million Palestinian citizens in Israel who have lived in the state since its establishment in 1948. They are usually referred to as 'Israeli Arabs' or 'Israeli-Palestinians'. This population perceives the Jewish State as an undemocratic and oppressive system that is unable to provide them with a source of identification. The national hegemony privileges the Jews in Israel with the collective rights of a nation, while the Palestinians are entitled to benefit only from their rights as individual citizens. Israel, as a state that was created first and foremost by and for the Jews, defines the Palestinians as religious minorities (Muslim, Christian, Druze and Circassian) outside the Jewish, hegemonic collective.

After the war in 1967 and the Israeli occupation of the West Bank and Gaza, the physical, emotional and intellectual links between Palestinians inside and outside Israel, which had been severed in

1948, were renewed. The identification of Israeli-Palestinians with the formation of Palestinian nationality and their support for its movements have generated the construction of their identity as 'Israeli-Palestinians'. The intellectual Israeli-Palestinians play an important role as cultural mediators who can explain Israeli political culture to the Palestinians and other Arabs, and at the same time disseminate Palestinian identity and culture within Israeli society (Kimmerling, 2004).

A look at Israeli society, especially since 1967, reveals extensive changes. In relation to the Israeli-Arab conflict, peace agreements with Egypt and Jordan have been signed, as has a mutual recognition agreement with the PLO. On the other hand, there have been wars in Lebanon and Iraq and two Palestinian uprisings (*intifada*) in the Occupied Territories. In relation to interior affairs, the Labour Party's hegemony came to an end, shaking up the secular, middle class and Westernized 'Israeliness'. This has been followed by a more open, multicultural rhetoric that has encouraged ethnic and national co-communities to fight for articulation, visibility and representation in both the symbolic and political spheres.

Israeli-Palestinians have undergone a process of modernization in education and economics as well as deep changes in their political behaviour. This includes changes in their voting patterns, the development of national organizations and political parties, and an intensifying process of identity formation. Although Israeli society seeks to become more democratic and partially enables the moving of co-communities into the centre, it still excludes its Palestinian citizens from this mobility (First and Avraham, 2004).

Two uprisings by Israeli-Palestinians signal the process of construction of their national identity and their dissidence in regard to constant discrimination by the Israeli establishment. Land Day, on 29–30 March 1976, constituted a series of violent protests against the government plan to appropriate land in the Galilee belonging to Palestinian citizens. Six Palestinians died in confrontations with the forces of government and many others were injured. Since then, this day has become a day of commemoration among the Palestinian co-community, symbolizing their united struggle for social, economic and cultural rights. The second crucial event is related to the Israeli-Palestinians' role in the Al-Aksa *intifada*. On 1 October 2000, the first *intifada* in the Occupied Territories erupted, following the visit by then opposition leader Ariel Sharon to the Dome of the Rock in Jerusalem. The Israeli-Palestinians, supporting their brothers, responded with angry protests that developed into violent clashes with the Israeli security forces (First and Avraham, 2003).

These confrontations continued for 10 days in several places throughout the country, leaving 13 Palestinian citizens dead and many others injured. As noted by First and Avraham, these two events indicate that 'While in the late 1980s Israeli-Palestinians were more involved in the civil arena, and their identity as Palestinians was just emerging, in 2000 their identification as Palestinians became their main cause, along with their civil struggle for equality and recognition within the State of Israel' (First and Avraham, 2004, p. 62).

Amal Jamal, a political science commentator, further conceptualizes this process, showing how the Palestinian citizens of Israel have recently increased their political mobility, 'interlacing social justice, distributive equality and national-cultural recognition into a political formula that could bring about a meaningful change in the lives of the Arab community in a state with a Jewish majority' (Jamal, 2007, p. 472). In other words, demands are being made that the state recognize its Palestinian citizens as an indigenous national group with collective rights beyond those of their individual civil rights. Jamal defines this complex relationship as 'the dialectics of state-minority relations' (473), claiming that Israel, as a nationalizing Jewish state, adopts political, economic and cultural policies that 'hollow out' (472) the citizenship of Israeli-Palestinians as a national, indigenous community.

This national identity construction increases the contradictory and ambiguous existential experience of the Israeli-Palestinians. They have neither fully collaborated in the renewed national culture of the Occupied Territories and of the Palestinian Diaspora, nor have they yet entirely integrated into the Israeli socio-cultural reality. They thus suffer from a constant double marginality, positioned on the periphery of both Israeli society and the national Palestinian movement (Kimmerling, 2004). David Grossman, an Israeli Jewish poet and author, describes Palestinian citizens in this problematic state as 'present-absentees', meaning that the Jewish consciousness perceives them 'as a group, which does exist, but is lacking face or names, one of homogeneous features, most of which are negative' (Grossman, 1992, p. 226).

Moreover, the Israeli-Palestinian conflict, which is an ongoing and central dispute, seems currently to have reached an impasse. 'A society that lives in a constant intractable conflict', notes Daniel Bar-Tal, a social psychologist, 'experiences insecurity, pain, bereavement, helplessness and depression. In order to confront such a difficult situation, the citizens need to understand the conflict, to obtain a positive identity and mental power to successfully stand in front of the opponent' (Bar-Tal, 2006, p. 17). Each side of such an intractable conflict develops a psychological repertoire constructed upon three central components. The first is the ethos of the conflict, which contains those mutual beliefs that help to signify the present and to indicate the future direction for the society. The second is the collective memory, by which the society narrates how and why the conflict began and developed and what were its constitutive events. The third component in the psychological repertoire is that of emotional orientation such as fear, anger and hatred toward the opponent. This repertoire, as Bar-Tal explains, helps each side to justify itself and negate the other, and consequently blocks any potential for a peaceful resolution to the conflict.

Looking at the conflictual relationships between Jews and Palestinians within the State of Israel, it is clear that the hegemonic process of socialization into conflict through factors such as schooling, army service and mass media and mass culture, seeks also to socialize the Palestinian citizens despite still keeping them on the margins.

Although positive changes in regard to civil rights may only just be creeping into view, they do exist, while the accumulating, assertive claims of the Israeli-Palestinians, which also ride roughshod over national collective rights, endanger and may serve to hinder any positive progress. The Jewish collective subconsciousness perceives the Israeli-Palestinians as inseparable from the entity of 'the Arabs as the total enemy', or 'the total Other'. The collective sociopolitical activity of the Israeli-Palestinians thus reinforces the mythic suspicion and anxiety of the Jews. Many Jews in Israel feel that the struggle over the validity of the Jewish State is not yet over, and that its Palestinian citizens are not a minority but part of a large majority that constantly threatens to undermine the State of Israel.

This contextual introduction highlights the difficult situation of Israeli-Palestinians living in a state of limbo between Israel and the national Palestinian agenda. It also indicates the ambivalent perception of the Palestinians by the Jews, who have not yet decided whether to include or exclude the Palestinian citizens of the state.

Ramla,[1] one of the mixed towns in Israel, is located about 15 km east of Tel Aviv. It was founded at the beginning of the eighth century CE by the Umayyad Calif Suleiman Ibn Abd El-Malik. The town was built on sand dunes (the Arabic word 'raml' means sands). Jewish settlement there began in the ninth century and especially flourished during the tenth and eleventh centuries. Around the year 1150 the Crusaders conquered Ramla and constructed a magnificent church, which still stands today. In 1033 and 1067 two earthquakes destroyed the town, killing many people. In the fourteenth century the town became the capital of the Mamluks who built the 'White Mosque', a minaret and cisterns for storage of water, all of which still remain intact. From this point until the sixteenth century both Jews and Arabs suffered, with the town being conquered at various times by different nations. In the sixteenth century Turkish Ottoman forces led by Sultan Salim the First conquered the country, including Ramla. The Ottoman regime lasted until the British Mandate in 1917.

Jewish settlement was renewed in 1886, but it was only after the declaration of the establishment of the State of Israel and the War of Independence in 1948 that Ramla became a Hebrew city, mainly for Jewish Mizrahi immigrants. Today, 18 per cent of its 75,000 residents are Muslim, remnants of the great Muslim settlement that had existed in Ramla prior to 1948. Most of them live in the old city and earn their living as hired labourers, storekeepers, restaurant owners and agricultural workers. The crime rate in Ramla ranks, unfortunately, among the worst five cities in Israel.

Here? Now? To Love? (1998)

This intricate socio-political context provided the background for both Jews and Palestinians in Ramla who wished to turn to theatre. The theatre project was deliberately facilitated by a female Jewish student, Oshrat Mizrahi-Shapira, and a male Palestinian student, Yussuf Swaid.[2] The supervisor of the project, Rimona Lappin, noted that: 'In a mixed town I wanted to organize a mixed group and therefore I sent Yussuf to the Arabic neighborhoods and Oshrat to

the Jewish neighborhoods to recruit both Jewish and Arabic participants' (Lappin, 2006). Swaid recalls a successful encounter: 'It was
easy for me to connect with them, knowing Arabic created closeness
and a common denominator between us' (Swaid, 2006). The facilitators provided the inhabitants with invitations in Hebrew and Arabic
to join the foundation event of the community-based theatre. The
mixed crowd that had accepted the invitation enjoyed an active,
playful event that comprised various theatre games and a basic
presentation of the suggested theatre project. 'That evening', says
Lappin, 'we didn't mention the conflict at all. We preferred to leave it
to the internal process that we would facilitate with those who would
choose to join the theatre project' (Lappin, 2006).

The group eventually comprised 12 participants: six Palestinians
aged around 20 and six Jews aged around 40. Swaid was not surprised
that only young Palestinians had decided to join the project. 'From
my experience', he says, 'the adults are too desperate, without any
motivation. The idea to set up a theatre seems totally redundant.
They say "what will it give us?" Those young Arabs who did come
had all been already involved in the community in one way or
another. Each of them persuaded another friend to join the group.
This was how the system worked on the Arab side. What surprised
me was the fact that young Jews didn't come' (Swaid, 2006).

The facilitators, at the supervisor's suggestion, decided that in
the first stage of the creative process they would circumvent direct
articulation of the conflict and instead direct an expression of the
common denominator. 'We decided to avoid explosive materials and
to focus on handing them the basic theatre principles. We believed
that in time, as the group formation and crystallization progressed, it
would be able to contain the conflict' (Lappin, 2006). The ideological
and practical point of departure was that of focusing on how sym-
bolic-artistic activity such as theatre could build up personal-human
contacts that would serve as a sufficiently solid basis to then touch
upon the conflict. The facilitators thus anticipated that the future
performance would not only reflect the socio-aesthetic process but
would also advance and deepen the political process.

Mizrahi-Shapira, the Jewish facilitator, recalls that she came to
Ramla full of trepidation. 'It was my first time working with a mixed
Jewish-Arab group. I didn't know then how to approach the con-
flict. Soon I found out that the participants would not talk directly

about it either. They persistently insisted that "we live together" and
wanted to put on a happy show' (Mizrahi-Shapira, 2006).

Swaid, the Palestinian facilitator, provides an important insight:

> I can understand their resistance to deal with the conflict, so do I.
> I cannot deal with it all the time. There are moments when being
> occupied with my identity becomes repulsive. But what surprised
> me was that they completely rejected the opportunity to deal with
> it. What became mostly important to the Jews and the Arabs as
> well was to put the conflict totally aside. I felt they were afraid to
> shake or break some kind of a silent agreement between them,
> according to which the Arabs a priori accept their lower position
> in relation to the Jews. But then all of them, Jews and Arabs,
> strove to create a peaceful atmosphere as if telling us through
> their deliberate silence that they didn't believe there could be any
> solution.
>
> (Swaid, 2006)

From the beginning the Jewish participants brought up personal
and individual life materials, while the Palestinians, as is usual
among subalterns, brought up life materials in which the personal
was always collective and political. This of course is a crucial point,
revealing that the mutual declaration of the group to circumvent the
conflict was only partially successful. Living in a mixed city, where
the physical and mental presence of the prolonged and seemingly
intractable conflict overshadows every aspect of daily life, the personal is always political and vice versa.

One of the Jewish participants chose to tell her personal experience
as a single mother. She was bothered by her social and economic
status and wanted to elaborate this issue through the theatre. As she
unfolded her story, she mentioned in passing that she lived in an
old, abandoned Arab house. This detail – marginal from her point
of view – effectively expresses the Israeli hegemonic socialization
into conflict. According to this narrative, part of the Palestinian
population chose to flee, abandoning their houses, during the War
of Independence in 1948. From the point of view of the Palestinian
participants, however, this apparently small detail was immediately
crucial, as according to their perception of the conflict the 1948 war
was their *nakba* (in Arabic: disaster, catastrophe). The Palestinian

population was forcibly deported, leaving behind their homes, property and land. Thus, the expression 'abandoned house' immediately irritated the Palestinian participants and elicited the struggle over narrative which both parts of the group had striven to ignore.

Another personal story, this time told by a Palestinian participant, focused on her position as a single woman in Israeli society in general and her own co-community in particular. As her story progressed, she went into more detail about her parents forcing her to marry against her will and about the horrendous possibility that some male relative might kill her because of family 'honour'.[3] This story facilitated the exposure of the issue of family 'honour-killing' as a crucial internal problem, especially for the young within the Palestinian co-community. The Jewish participants, who were only partially acquainted with this issue through media reports, got the opportunity for a close encounter with this Arabic tradition and were able to see for themselves how contested and problematic this issue was in the lives of the younger Palestinian generation. The Palestinian participants discovered that this issue interested their Jewish colleagues and elicited support for the Palestinian's social agenda. Additionally, the group created a link between the murder of women in the Palestinian co-community and the battering of women in the Jewish community, on the ground of the feminist stand against female oppression. This development of a feminist common ground and interpretation was one unique outcome of this group's making do with theatre.

The play, *Here? Now? To Love?* was written collectively by the group with the help of the facilitators. The basic narrative structure was that of unrequited love between a Jewish woman and a Palestinian man, into which parallel scenes between a Jewish couple and a Palestinian couple were inserted alternately. It is interesting to see the ways in which the narrative reflects those conflictual components that rose to the surface through the creative process. The mixed-race couple meets again after many years apart; in the intervening years she has become a single parent, divorced from a violent husband, and he is a father and is now a widower, following the death of his wife, who had married him at the order of her parents. For a while, it seems as if now that both are free of family ties, they will be able to consummate their love. Unfortunately, they soon realize that the national conflict infiltrates even the most personal-individual sphere and suffocates any possibility for reunion. Tamar, the poor Jewish single mother, who cannot afford a modern apartment in the new neighbourhoods in Ramla, lives with her children in the old city, in an old house that once belonged to the uncle of Amir, her Arab lover. Amir, in fact, had returned to Ramla in order to reclaim his inheritance.

The symbolic text thus publicly illuminated the contested issue of the Law of Return,[4] no matter how strongly the participants had tried to avoid the conflict. Moreover, the theatrical event posited a particular case that created a fusion of the personal, the political, the social, the national, and past and present. This form of grounded aesthetics, which confronts the hegemonic compartmentalized regime that erects clear borders between the personal/local and the political/national, could have been created only by a mixed group of Jews and Palestinians whose daily lives are intertwined.

The hidden conflict was also implied in the *mise-en-scène* and scenography. Large modular cubes signified various locations, with the process of deconstruction and reconstruction being carried out in front of the audience by the Jewish and Palestinian performers interchangeably. 'This gestic passage from scene to scene', says Lappin, 'echoed the ongoing internal contest between the two groups on status and expression' (Lappin, 2006).

The production process of the play was tense but the mutual effort to complete the creative process and stage a performance propelled the group forward. This shared, practical and immediate goal enabled the ensemble to present on stage a utopian closure, expressed through a performance of a wedding ceremony that was neither Jewish nor Arabic but at the same time not-not Jewish and not-not Arabic. The mixed loving couple were both acted by Jewish performers, who were of a more appropriate age to play the older characters. This of course was the formal, functional explanation, but I would suggest that it was also a poaching (subversive) tactic to present an alternative model in a way that would be acceptable to the mixed audience of Ramla. Was the wedding a manifestation of deus ex machina? Was the union on stage that of the man and the woman as performers? As characters? Was the theatrical closure a suggestion for a new, mixed Jewish-Arabic ceremony? Or was it a symbol for a new, totally alternative social order?

These were actually the questions that began to be raised by the mixed audience in the discussions after the performances. However, the cultural centre's authorities cut the discussions short, giving technical excuses, and thereby exposing once more the establishment's deeply ambivalent feelings towards theatre in co-communities: the fear that making do with theatre might shake the status quo not only weakens but at times completely blocks the theatrical energy of the co-community to instigate change.

Who Killed Achmed Chamed? (1999)

Discussing the tense relations within the mixed group, the facilitators decided that each community would work by itself for a year and then the two groups would reunite as one ensemble. Accordingly, Swaid facilitated the new Palestinian youth group, which comprised several veterans but mostly newcomers from Juarish, an impoverished Arab neighbourhood, notorious for its high rate of violence and drug trade.

'This theatre project', Swaid clarified, 'was immensely worthwhile for the Arabs, as it provided them with the chance to concentrate on themselves, on their lives within their own community and to bring up those particular internal problems that mostly bothered them. It was highly significant for them to put on a show in Arabic that would reflect the internal quarrels between the leading families (*chamula*)' (Swaid, 2006). The previous theatre activity had been an important preparatory stage but, nevertheless, the young Palestinians needed an autonomous protected space within which to find their own voice and articulate it through theatre.

The play *Who Killed Achmed Chamed?* (1999), was a self-reflexive, critical account by the Palestinian youth of the level of violence in their neighbourhood. The performance was deliberately staged outdoors, near the ruins of the old Muslim cemetery, the 'White Mosque' and the minaret. The play was constructed as a court trial in which various witnesses gave their testimony. The movements of the performers' feet on the bare ground produced clouds of dust that provided an effective setting, physically and visually expressing the violent atmosphere of the neighbourhood. These poaching tactics created a stage metaphor that delivered two interwoven messages. The focused message articulated the social agenda of the youngsters seeking to bring about change within the traditional power relations in their own community. The real historical background suggested the glorious days of ancient Muslim Ramla, providing the second message, about the complex identity construction of these young people.

Ramla's Beach (2000)

A year later, a mixed Palestinian-Jewish group was reorganized, and together with Swaid and Mizrahi-Shapira, put on the performance *Ramla's Beach*. The narrative focuses on a lifeguard's cabin where various miserable people come to commit suicide, following the instructions of the lifeguard. These characters were intentionally classified not by ethno-national categories but by broader and more open social categories such as 'addicts', 'former addicts' and 'women', who could be acted by either Palestinian or Jewish performers. 'The idea was', says Mizrahi-Shapira, 'that all the characters are between life and death, moving toward their death, while from time to time a relative breaks in and tries unsuccessfully to save someone from committing suicide' (Mizrahi-Shapira, 2006).

Ramla's Beach was a dark and horrifying allegory of the mutual oppositional stand of the mixed group in relation to the establishment, which constantly neglects Ramla and sentences to lives of desperation both its Jewish and its Palestinian inhabitants. However, in the discussions with the local audience, the theatre group discovered that it was the Arab spectators who were the most shocked by the performance, and that they were primarily affected by its critical message on the issue of family honour-killing. For the Palestinian co-community then, the secondary message became the primary one and of highest importance.

This audience response revealed two significant social facts that had neither been presented in the public sphere nor had found their way into academic critical socio-political discourse. The Palestinian co-community is experiencing a complex internal struggle over what is the right way to live, particularly in light of the Islamic fundamentalism that has reached the Israeli-Palestinian communities. However, the internal dispute over family honour-killing that was exposed by all three performances – *Here? Now? To Love?*, *Who Killed Achmed Chamed?* and *Ramla's Beach* – is a sign of a much broader inner struggle, especially between the young and the old, over Arabic tradition, history and memory. It seems that the unsettling question of ideological and practical anchors to their daily life as a community bothers the Palestinian no less, and perhaps even more, than the Israeli-Palestinian conflict. The performances thus provided an emic, thick description[5] of the gaze from below, which reveals a co-community in a dynamic, turbulent process of passage. Israeli-Palestinian intellectuals such as Amal Jamal, Azmi Bashara and Achmed Tibbi, tend to emphasize and reinforce the discourse on the emergent national identity of the Israeli-Palestinians, while silencing the internal, local conflicts, which are also an integral part of this identity construction.

Practising theatre not only provided the Palestinian performers with a protected space in which to give voice to an otherwise muted issue but also 'contributed to provoking change', as Swaid indicates. 'When we first asked them whether they would kill their own sister they replied that if she would shame them they would have definitely killed her. Later on, during process, they talked differently, apologizing that they had no choice as this is a traditional duty that forces them to do so even if they do not want to' (Swaid, 2006).

In the performance the responsibility for the murder is projected onto the mother who puts the gun in her son's hand and sends him out to fulfil his 'duty' and kill his sister. 'Here', says Swaid, 'the internal criticizing of the mother was most important, since we wanted to show that when a daughter transgresses, it is the mother, and not the father, who feels dishonoured and seeks to avenge the insult' (Swaid, 2006). This exposure of the mother as the chief bearer of a tradition that actually oppresses her, was another unique insight raised by this performance that called for attention and further discussions. Studying community-based theatre in Israel reveals that from the 1990s on this cultural practice has been characterized as a subversive device that enables a co-community to become self-reflexive and self-critical. Groups of Mizrahi women in particular used the theatre to articulate the conflictual daughter-mother relationships in their co-community, highlighting an oppressed mother who herself behaves as an oppressor.[6] This insight into the cycle of oppression in the domestic sphere demonstrates the complex and paradoxical structure of feeling of the subaltern.

Focusing on the critical subject of family honour-killing, which related to the Arab community, the audience consciously or unconsciously bypassed the shared articulation of the performers of their mutual feelings of deprivation as two co-communities in Ramla. This unique message, which the audience seemed to neglect, points in fact to the bridging power of facilitating theatre in places of conflict. 'Making do with theatre' eventually generated a dialogic and bridging process through which the mixed group managed to reach a common ground. The insight reached, that Jews and Palestinians in Ramla are in fact both co-communities and should articulate and confront their marginalization together, was a clear and unusual outcome not only of the third performance but also of the extended theatrical event that comprised all three creative processes and performances. This genuine, subversive insight had indeed required a long process. At the beginning it was the sheer amount of effort on all sides to produce a show that connected the Jewish and the Palestinian performers. With time, the process of making theatre created a dialogic, bridging process that eventually produced the third performance as a mutual articulation of the group as a mixed co-community.

The Rich Man Dies from Laughter (2001)

A year later the mixed group wrote and produced the play *The Rich Man Dies from Laughter*. This was a comedy about a family that has no choice but to learn to live together in peace in order to receive its inheritance.

When I told Mizrahi-Shapira that I perceived it as yet another allegory of the national-political conflict, I immediately reawakened her confusion and discomfort: 'I wanted to discuss with them the conflict. I was miserable that it was stuck. It remained in the background' (Mizrahi-Shapira, 2006). Swaid, her Palestinian partner, did not see eye to eye with her, re-exposing the implied conflict between the two facilitators too. 'I'm against opening the conflict by force. I completely understand the group's silence, their profound fear' (Swaid, 2006). This statement places the whole theatrical event in Ramla in a somewhat pessimistic light. Nevertheless, it is important to listen to him. Making do with theatre is by no means merely for articulation – to enable the inarticulate to find their own voice. The Jewish-Palestinian theatre group from Ramla deliberately used silence and muteness as a poaching tactic not only for survival but also and mostly as a form of a high-pitched, hopeless cry that outsiders can barely hear. When making do with theatre might be dangerous, circumventing the conflict is interpretable as another tactic of articulation.

'And what about the prospect of undoing the conflict?' I asked Swaid. He replied:

> I myself have been lately involved as a professional actor in both institutional theatre and television. My art as an actor is not an escape. Art stimulates a more profound and interesting form of discussion than that in daily life. In my art as an actor I create and provide this phase of pure listening for the audience, which is very important. But at the same time I also love my work as a theatre facilitator with Arabic youth groups. While Arab culture is still being marginalized in Israel, community-based theatre con-tributes to change this situation. In Ramla, for example, a new Arabic cultural centre has been opened as a direct outcome of our theatre activity. Community-based theatre indeed provides the Arabs with their own autonomous space to find their own voice. I think I'm the only Arabic artist who studied community-based theatre at the university. I know how to reach the Arab youth. I understand the mind of the minority.
>
> (Swaid, 2006)

The Jerusalem Jewish-Arab theatre project

In 2002 Swaid left Ramla to continue his acting career although in his free time he still did theatrical work with young Arabs in Jaffa. Oshrat Mizrahi-Shapira tried to carry on the theatre project in Ramla with a new Arab co-facilitator, Sa'ad Tali, a social worker who was also qualified as a group coordinator and mediator. Tali suggested a new method of work that focused on a direct confrontation with the political conflict. In this model the theatre functioned as a means to stimulate dialogue in order to study Israeli-Palestinian relationships not only in Ramla but also on a national scale.

Mizrahi-Shapira hoped that the new work process would engender a change in the group's behavioural pattern, but she was disap-pointed: 'They were stuck to their desire to practice theatre and to leave politics behind' (Mizrahi-Shapira, 2006). The new method proved to be too difficult and after a few months the group decided to disband.

Mizrahi-Shapira decided to continue working with Tali and looked for a new mixed group. She noted that: 'From a very young age I have participated in meetings between Jews and Arabs. Perhaps this is because of my own feelings as a subaltern, being brought up in Katamon, an impoverished Mizrahi neighborhood in Jerusalem' (Mizrahi-Shapira, 2006). She and Tali moved to Jerusalem, where they continued to facilitate theatre with a mixed group for four years. This latter project was devised by the 'Theatre as a Mediation Tool' workshop co-sponsored by the Billy Crystal Project: Peace through the Performing Arts, of the Theatre Studies Department at the Hebrew University and the Khan Theatre.

Tali notes that he and Mizrahi-Shapira sought to provide people with a better understanding of the roots of the conflict. 'The workshop is intended for those Jews and Arabs who want to confront themselves and "the enemy" through a strong experience' (Tali, 2006). The premise of the work process was that in order to undo the national conflict it was also important to take into account the identities of individuals and the psychological components of these identities.

Accordingly, the process was planned in three stages: the personal connection, the group connection and the national connection, with each stage forming the basis and becoming the generator for the following stage. Tali explains that the personal starting point is crucial, as 'the conflict also stems from the universal mental structure that stimulates us to project onto the Other our own problems and angers' (Tali, 2006).

The political aim of the project was made clear as soon as participants were recruited. The facilitators emphasized that the mixed group was being organized in order to confront complex issues in relation to the conflict. 'The combination of group dynamics and theatre', notes Tali, 'offered the optimal system to release heavy and blocked moments. It created an inner balance between those participants who liked to talk and those who preferred to express themselves through playing and acting' (Tali, 2006). The gender, national and professional encounter between Mizrahi-Shapira and Tali, which successfully unfolded in front of the participants, functioned as a model relationship for them.

The group comprised Jewish leftist students, mostly Ashkenazim, Muslim and Christian Palestinian students from northern Israel, and Palestinian inhabitants of East Jerusalem. The Palestinians were the hardest to bring into the project as they often expressed clear disbelief in the ability of such a project to produce change. As the political reality seemed to be becoming increasingly intractable, the Palestinian students became more pessimistic and antagonistic to any kind of Israeli establishment initiative. Moreover, they were afraid of being identified as collaborators and traitors by their community. It is important to note that once the group was complete, Palestinian women outnumbered the Palestinian men. This contextual asymmetry between Jewish and Palestinian recruits was balanced at the beginning of the project by the warm and empathetic welcome extended to the Palestinian participants by both facilitators and the Jewish participants.

The project lasted for several months and was to culminate in a theatre performance, signalling the end of the process for the participants and the beginning of the reflexive process for the audience. Looking back, Mizrahi-Shapira observes that 'the concrete task to put on a show, as professional as possible, encouraged and united the Jewish and the Arab participants. Their commitment to the same mission was exciting for them and for us as well' (Mizrahi-Shapira, 2006). This form of coalition, which had also characterized the Ramla group, foregrounds once again the multi-beneficial potential of making do with theatre, no matter what kind of working model one chooses.

The encounters in the Jerusalem project developed a steady ritualistic structure. They usually opened with a short round of personal associations, in which each participant expressed their immediate feelings. This was followed with a few warm-up activities, drawing the group into the central theatre exercise around a concrete political component. The final phase was the sharing section, in which the group discussed the encounter.

The facilitators planned a process that was to start with a short phase of group consolidation and be followed by a longer phase in which the two national groups would work separately to confront the conflict; finally, the two groups would be brought together to produce the show. However, it quickly became clear that the initial phase of acquaintance and consolidation needed more time, since both Jewish and Arab participants were at first over-polite, trying as far as possible to repress the conflict. 'They were regressive and frightened', Tali recalls, 'They were well aware of the kind of a project to which they had come, seeking to immediately present a "new Middle East" within the group' (Tali, 2006).

The following excerpt, written by Shai Klaper (Jewish, male participant), critically reflects the suspicious and false atmosphere in the early encounters:

<u>A Cherry</u>
I fancy licking something
I fancy licking something like an ice cream cone
Yes, sit on the soft sand on the beach and lick ice cream
Politically correct flavored ice cream
To lick the white part
And the black
The strong layers of the biscuit
The soft layers of the cream
To lick all the colored sweets.
To lick it all in gentle pluralistic movement of the tongue
And spread on top some humane sweet whipped cream.
Spread on the mouth,
On the lips
On the tongue
Without stopping.
Sprinkle some rice puffs on the poor
So I will not be able to see.
And stuff glazed cherries in the ears
So I won't be able to hear
Fraternity ice cream, peace ice cream, justice ice cream
Oh boy, I'm melting![7]

One of the exercises that aimed to promote personal acquaintance was that of the identity play, in which the participants had to adopt one another's identity. They were instructed to move around the room, stopping to face one another. At each encounter each participant introduced himself/herself to the other by using the identity of his/her previous partner as if it was his/her own true identity.

The various identities thus circulated within the group until each participant came face to face with his/her own identity performed by another participant. This exercise had an immediate effect on the participants since it demonstrated the elusive and dynamic quality of identity. It also revealed identity as a social construct that may differentiate humankind by gender/nationality/class/ethnicity, but within which individuals are more capable of negotiating than they might have thought. This exercise was followed by a more complex outdoor activity: the group was divided into mixed pairs and each individual took his/her partner to a significant place and recounted a piece of his/her biography in relation to that place. This activity stimulated face-to-face interaction and a flow of gentler information that deepened the interpersonal relationships and facilitated a challenging of stereotypes and stigma on both sides.

Only after the facilitators felt that the participants had established a sufficient degree of trust and group consolidation did they move to the second phase. Each national group then performed in front of the other, presenting first their image of the political reality as-is and then an alternative image designed to change and improve the perceived reality.

This exercise was followed by symbolic activity around objects as indexical signs of 'home'. Each participant introduced a personal object and told its story. Then, each national group created a theatrical scene that included all the objects of that group. The Jewish 'home' tended to be multivocal and more universal, while the Palestinian 'home' was more traditional and ethno-specific. Moreover, the Palestinian 'home' created a metaphor for the Arab village before 1948. This image work exposed the cultural, emotional and intellectual distance between the Jewish, Westernized, liberal group, whose ideology was based on individualism, and the more traditional Palestinian co-community, whose personal experience is usually concomitantly collective and national. Accordingly, the facilitators then directed the Jewish group to seek their national identity and the Palestinian group to seek their individual, personal identity.

This stage, when each group had performed in front of the other, aroused antagonistic responses from both sides. The Palestinian sub-group expressed sharp accusation: 'You – they did to you the Holocaust and you – did a Holocaust to us.' The Jewish sub-group, with feelings of guilt mixed with a wish for acceptance, counter-claimed: 'We may be guilty but nevertheless we are here, so what? Will you throw us into the sea?' (Tali and Mizrahi-Shapira, 2005).

The facilitators encouraged both sub-groups to vent their suppressed feelings of rage, fear, hurt and anxiety. The therapeutic assumption was that only after cleansing the deep negative feelings that are usually bypassed, would it be possible to arrive at the next stage of reciprocal listening.

The following short scene was presented by a female Palestinian performer, Rabbav Badran, and a female Jewish performer, Ori Goldstein, who created a single condensed image of the constant struggle:

<u>Heads or Tails</u>
The performers stand tied with their hands bound together with a red scarf. The Jewish woman moves impatiently, looking at her watch while the Arab woman is napping. In the background is the sound of a siren signalling on set of the memorial day for the Israeli dead soldiers. The Jewish woman immediately stands erect.
Arab: (tries to pull her hand down) stop it, it hurts.
Jewish: Why didn't you stand up?
Arab: Why did I have to stand up?
Jewish: In this house we stand up to honor the kettle every day! Why do you smile? You are doing it on purpose, I know.
Arab: Before the kettle there were other instruments here, but with you, all the time is the kettle!
Jewish: Let me remind you that you are renting this place so you must obey my rules. If you don't like them, please look for another apartment!
Arab: You may think that I rented this house but actually this house belongs to me as well!
(They start to physically fight)
Arab: Let's rise up!
Jewish: Let's be astonished!
Arab: Let's attack!
Jewish: Let's defend...attack!
Arab: Let's defend!
Jewish: Let's conquer!
Arab: Let's evacuate!
Jewish: Let's exterminate!
Arab: Let's destroy!

(The performers turn their backs on each other while an explosion is heard in the background)
Arab: Do you hurt?
Jewish: A little, and you?
(They approach the audience in a mixture of Hebrew, Arabic, English and French) Can somebody help?
Arab: (as if hearing distance voices) Sho? Shalom?
The Jewish: Shalom?!
They take off the scarf, drape it around their necks, smile and then hang themselves.

Goldstein and Badran became close friends through the project and continued to perform this piece at various events later on.

This humorous-horrifying domestic allegory exemplified the unique form of grounded aesthetics that this theatrical event was able to present. It also clearly demonstrated the power of making do with theatre to generate a double effect of political therapy. The first effect lies in the mutual artistic formation and articulation of deeply suppressed feelings, while the second effect lies in the mutual ability to contain such a cruel articulation. This stage in the work process enabled the creation and reception of the self-text produced by the Palestinian female participant, Houda Otman:

Daddy's complex
Eight times one is eight, eight times two is sixteen, eight times three is twenty-four, eight times four thirty-two, eight times five forty, eight times six – eight times six ... My father cannot tolerate or hear the number forty-eight. Do you know why?
Mum once said to Dad: 'Abu Ali, I got you trousers, size 48. Do they suit you?' Do you know what happened? He grabbed the pants and threw them out the window! I cannot forget that day! I ran to the window looked down and saw someone taking the pants. Ever since that time, my dad lies in bed without pants, naked. It is forbidden to wear pants size 48. Forbidden to wear shoes size 48. Forbidden to celebrate his 48th birthday, this year! Liesh ('What'? in Arabic)? 'Why does Dad behave like this?' I asked the doctors. One day I looked and found his diary.
She reads from the diary: The year 1948, the place Dir Yassin. The time – a disaster. That day mama was hugging me and the hug started

to move away from me. Mama, Mama, what are you holding in
your hand? Mama, who are the strangers? Why are there dead bodies, what are you holding? You are hugging a pillow instead of me!
I started to cry, my tears rose from one Arab village, which was
damaged by the Israelis, to the next. (He listed the names of the
villages in his diary, the translator explained to us through her
narrative.)
Since then my father sleeps without pants and without a pillow.

This monologue was received with strong applause and stimulated
great excitement among the Jewish participants. Houda, it is important to note, had hardly ever spoken before, and did not appear to
be very eloquent at first. Nevertheless, in this stage of the process,
this almost mute young woman finally felt safe enough not only to
speak out in a loud voice but also to articulate a traumatic issue. Her
poaching tactic of using the number 48 stimulated a sense of reception and understanding for a Palestinian girl who talked haltingly
in her own language about the Palestinian villages that had been
destroyed.

This reception of the Arabic narrative by the Jewish participants
was balanced in the scene that followed. A very quiet Jewish participant came onto the stage, presenting a religious woman wearing a
white dress and with her head covered with a scarf. She started to
sing the national anthem of the State of Israel. Her face expressed
the endless journey of the Jews as she sang the pain and the Jewish
distress, for *Hatikva* (The Hope), the anthem, contains more helplessness and pain than hope. This piece was excitedly received by
the Palestinian participants, who confessed that it was the first
time they had ever listened to the Israeli anthem without blocking
their ears.

When I asked Mizrahi-Shapira and Tali about their own personal
processes as Jewish/Palestinian and female/male co-facilitators, she
said:

I understood from my own experience that as a Jew I suffer real,
existential anxiety. When we started the project, I felt guilty, as
the Jewish participants often felt. I wished to give, to contain
and at the same time became more frightened. I recognized this
ambivalence in myself through the work process. I also perceived

the complexity and the basic need of both sides to exist and how difficult it is to attain without threatening the other.

(Mizrahi-Shapira, 2006)

However, as part of the Mizrahi co-community, she could not help feeling an empathy toward the Palestinian co-community, perceiving the deep, hidden link that transgresses the national-political conflict. 'I was taught that it is not tactful to speak about my discrimination, so I speak it through the Arab. In this sense we are on the same side' (Mizrahi-Shapira).

Tali answered that his approach stemmed from his profession as a social worker and group facilitator:

> Here, of course, I became more emotionally involved. It enabled me to deal with something that constantly bothers me – my identity, my life, my fears, and my existence. I think it very important that more Jews and Arabs attend this project. It facilitates a deep encounter with the Other. It stimulates a strong influence. I personally believe that there is symmetry in the distresses of both sides. These two nations are not bad in their nature. They have both suffered and want to live. I do believe that they can exist side by side. Here in the project I feel it, not only intellectually or theoretically.
>
> (Tali, 2006)

This project suggested a form of making do with theatre that handled the conflict on a small-scale, within a frame that was both protective and demanding. The insight that this theatrical event engendered was that one needed to attain a particular mental state to be able to contain the harsh articulation of the Other, and that this was the necessary basis on which change in the violent relationship could be brought about.

It is possible that such a small-scale theatrical model of conflict management can also be adapted for managing the conflict in the larger frame. The two mixed-group theatre projects in Ramla and Jerusalem reveal that there is a gap between the monolithic public discourse and theories about the Israeli-Palestinian conflict and the dense and contested grass-roots theatrical narratives. Practising theatre encourages ordinary Palestinian and Jewish citizens to strive for a mutual artistic creation that can symbolically reflect their reflexive and critical gaze inside and outside their own community, and inside and outside the mixed theatre group. The self-texts that these projects publicly posited were multifaceted images that dared to expose internal problems and existential anxieties. They were on the one hand more complex and informative than the leading political discourse of both Israeli and Palestinian speakers, while on the other hand they implied a modest possibility of finding common ground and a mutual wish to undo the conflict.

9
The National Festival
of the Co-Communities:
the Meta-Theatrical Event

Each chapter so far has presented a particular co-community and its ways of making do with theatre for articulation and empowerment. As an epilogue to this book, I have chosen to discuss a large-scale, spectacular theatrical event in which various co-communities came together in a central theatre venue in order to celebrate their national festival of theatre in co-communities.

Theatre in co-communities in Israel is currently a dynamic cultural practice engaging Mizrahi Jews, Ethiopian Jews, Jews from the former USSR, Israeli-Palestinians, the disabled, the elderly, prisoners and battered women. There are theatre groups that operate independently in various places throughout the country, and thus the explicit intentions of the national festival are to facilitate interaction between these groups, to discuss mutual practical issues, and to try to reach greater institutional recognition and to gain an increased budget for additional theatre projects (Alfi, 2002).

As all the theatre groups operate in some way on the geographical, social, economic and cultural margins, the concept of an annual national festival, in which different co-communities come together in an elegant theatre building in the centre of the country is a tactical manifestation of their joint move from the margins to the centre, and symbolically transforms their invisibility into an assertive visibility. This celebration implies that co-communities can make do with theatre in order to foster alliance, and that as a coalition they can demonstrate greater symbolic power to challenge the status quo. Thus, the super or meta-theatrical event of the festival reveals the potential of theatre in co-communities to generate articulation not only in its

161

political meaning of expressing muted, unarticulated voices from 'the bottom' but also in its second political meaning, that of binding people together into a cohesive social collective. Ernesto Laclau and Chantal Mouffe (2001), indicate that in contemporary Western society there is a need for a process of articulation between the different subject positions in order to construct a collective subject and identity. Co-communities have first to perceive their relationship with the hegemony as oppressive and only then can they be articulated into a cohesive, resistant collective. Theatre in co-communities, as I hope this book has revealed, is a symbolic political practice that binds the performers in a way that affects their identity formation. The creative process motivates individual and group reflexivity, raises social consciousness, and thus articulates the performers as a collective subject with a socio-cultural identity that makes do with theatre in order to articulate the voice of their particular co-community, which is usually also the voice of their larger co-culture.

The national theatre festival of the co-communities is a foregrounded display of articulation in both its political connotations.

Contextualizing the festival

The National Community Theatre Festival initiated by Yossi Alfi in 1998 took place annually until 2004.[1] Alfi, born to Jewish immigrants from Iraq, and now a well known director, actor and poet, was one of the founders of community-based theatre in Israel. Back in the 1970s he led a radical theatre group that set out to express the oppressed feelings of the Mizrahi ethnic group.[2] He later changed his orientation and left community-based theatre almost completely in favour of other more mainstream theatrical projects.[3] In 1998 he was appointed to administer the Givatayim Theatre, a new arts centre in greater Tel Aviv, where he decided to organize a national community-based theatre festival. It appears that Alfi, who began his career on the margins and then gained power in the centre, had not forgotten his origins, and eventually used his position to benefit those on the margins of whose work he had once been a part. 'Givatayin theatre', states Alfi, 'is an international arts centre that hosts theatre companies and art exhibitions from all over the world, music concerts and also the best theatre productions. The theatre is also located in a very affluent neighborhood, and so it is here, deliberately, that I want the national community-based theatre festival to take place' (Alfi, 2002). The meagre funding that Alfi managed to obtain from the Ministry of Education and the Ministry of Science, Culture and Sport necessitated a great deal of volunteer work, especially by Alfi's wife Sue, the producer of the festival, and by Levana, Alfi's secretary,[4] as well as by his young daughter, who acted as an usher at the event itself, and several leading community-based theatre activists – Alfi's acquaintances – who willingly participated whenever Alfi requested it.[5] This 'family-like' process of production contributed to the development of an especially close and bonding atmosphere of 'communitas' (see below), but it has also been criticized by those community-based theatre practitioners who feel that the festival suffered from being a one-man show, and that Alfi should have stopped acting as the sole authority whose final decisions determined which co-community was 'in' and which was 'out'. The festival usually lasted two days, from 4 p.m. on, and took place in the autumn. Various community performances, workshops and discussions fully occupied all the spaces of the theatre, including the large open foyer, stairs, yard and coffee shop in the front of the building. Each festival attempted to focus on a central theme. In 1999 it was 'Jews and Arabs', in 2000 it was 'women', in 2001 it was 'groups with special needs' and in 2002 it was dedicated to 'social community theatre' in general. Nevertheless, in practice, each festival articulated different voices from various places, and created a colourful, polyphonic image, which truly represented contemporary Israeli theatre in co-communities as a multiplex cultural phenomenon. Entrance was free of charge to the general public, but the audience nevertheless mainly comprised 'insiders' with direct connections to theatre in co-communities: the performers themselves, their followers, theatre students and practitioners, community workers, social workers and delegates from the municipality. In 2002, for example, Alfi sought to expand the number of 'outsiders' by arranging the grand opening of an art exhibition at the same time as the festival. Although this managed to pack the entrance to the building, the stairs and the lower floor where the paintings had been hung, it did not draw any additional spectators to the performances themselves. It was somewhat strange to see the physical proximity of these visitors to the festival audience while at the same time being aware of the political and emotional distance between them. Nevertheless, I believe that as a poaching tactic Alfi's idea was at least successful in adding some visual and auditory 'volume' to the whole festival event, which determinedly strove to imagine itself as more glamorous than perhaps it really was.

Any theatre group that wished to participate in the festival first had to submit a video cassette of its latest production with some basic information about itself. Selection of the performances was based on the criteria of social relevance of the subject matter and artistic quality. One or two of the best works received the privilege of being performed in full, while others were given partial presentations. Usually, a sequence was comprised of three or four pieces from different performances but with a common denominator. Such a structural system acted both to indicate only 'the best of' and as a functional device to draw more groups to the festival. As Alfi noted, 'it helps to fill the auditoriums' (Alfi, 2002). In 2002 two such sequences were performed each day, with full performances and workshops opening and closing each day. The total performances constituted one meta-theatrical event through a collage composition. The intertwining of the different performance sections created a particular type of a theatrical event that was unique to this festival.[6] Each section was removed from its original local context, where it had contained a specific significance for a particular co-community, and was brought together with the other chosen sections in a collage that built into a new sign system that connoted new meaning and experience.

My research journey into the meanings and experiences foregrounded by the festival was based on the 2002 festival as a case study. The materials that I analysed comprised my own participant-observations, the video of the festival and 70 questionnaires completed by spectators. In establishing my interpretation I also employed three analytical categories borrowed from the social sciences. The Israeli National Community Theatre Festival was basically a public event like any other festival. A 'public event', as Don Handleman states, is something undertaken by people so as to make more, less or other of themselves; thus it always 'does' something, affecting the social order. Handelman distinguishes between a mirroring public event, which presents the lived-in world as it is; a modelling public event, which presents an alternative to the lived-in world; and a representing public event, which offers propositions and counter propositions about the lived-in world, and raises questions, and perhaps doubts, concerning the dominant social forms (Handelman, 1990). This particular festival constituted a public event in which co-communities made do with theatre in order to make more of themselves and thus to confront the dominant social order. The meta-theatrical event of 2002 offered various moments of mirroring which were transformed by the performative energies into modelling scenes revealing the potential for change and empowerment through theatre.

Public events, especially aesthetic ones, carry the promise of what Victor Turner (1967) describes as 'communitas', a special bond that unites people over and above any formal social bonds. In communitas there are undifferentiated, equal, direct, I-Thou or Essential-We relationships that are more than just those of casual camaraderie or ordinary social life.[7] In the festival the fictional worlds represented on stage modelled forms of communitas that infiltrated into the 'here and now' of the event, enlarging the circle of communitas by creating a bond between the various groups that had gathered to participate in it.

A community, as Benedict Anderson suggests, is in fact always an 'imagined community'. In the mind of each of us there is an image of our community. This image is invented, created, but nonetheless influences us as if it were real. Thus, communities are to be distinguished not by their falsity/genuineness, but by the style in which they are imagined (Anderson, 1991). Experiences and meanings in the festival were generated by the tension between the real and the imagined. The festival was that public event which provided the co-communities with a play-space in which to imagine more of themselves, to imagine themselves as equal, powerful and dominant; and as theatre artists who managed to create relevant and emergent theatre in the centre of Israeli culture and society. This invented image of the festival bore psychological and political significance not only for the participating co-communities but also for the lived-in world outside, which continually dismisses theatre in co-communities as residual theatre.

Communicating experiences and meanings

The festival opened with a low intensity piece, in a small hall in the basement, with two clowns playing in pantomime with each other and with the audience. It gradually became clear that one was a professional medical clown while the other was actually a practising gynaecologist who had graduated long ago from the famous mime school of Jacques Lecoq. After their performance they sat in front of us, with their red noses still on, and presented their mutual agenda to develop medical clowning in the hospital. They explained that their performance represented only short demonstration of clown-playing with young patients before and after treatments, and then humorously responded to questions and comments from the audience. What had been experienced at first glance as a side-event with little reference to theatre in co-communities appeared retroactively as a proto-event that indicated some basic constituents of the festival. In the performance it was the clown who wore a doctor's gown over his colourful garments and occasionally held a doctor's bag, while the doctor chose to completely abandon all the signs of his social status and fully realize his repressed ego. Inside and outside the fictional world they behaved like real partners with a special bond, facilitating one another 'to make more of themselves'. The mute clown became the doctor-clown explaining his funny medical methods and recounting his personal experiences from the hospital. The serious doctor became the clown-doctor, revealing his additional artistic skills. We, the audience, who had shown from the beginning a high level of participation and attentiveness, ended the event with a simple but humorous chorus of baby voices taught and led by the doctor, which signalled our reception into and expansion of the clowns' communitas. Thus, *To Smile*, the first event in the festival was a modelling event for alternative approaches to social roles, human relationships, medical treatment and, most of all, to theatre, community and community-based theatre.

The vertical transition from the basement to the upper floor was physical, mental and visual, exposing us stair by stair to bright, noisy and vibrant high intensity happenings in three different locations. On the plaza by the main entrance community-based theatre students performed *The Centre of the Campus* in the middle of a square of chairs. This was their own project produced as part of their second-year curriculum and the original script told the story of different kinds of homelessness as a critical metaphor of Israeli society. It was directed by Igal Azarati, one of the more socially and politically committed directors in Israel, who is himself a graduate of the Community Theatre Unit of Tel Aviv University, and who currently runs the Arabic-Hebrew Theatre in Jaffa.[8] The students were very disappointed in this performance, as they felt that their work should not have been removed from its original location in the university, where it was performed in a huge tent and in front of a captive audience.

In the festival, without the tent, other accessories and proper audio accompaniments, but with a lot of noise from the adjacent coffee shop and many occasional onlookers strolling in and out of the building, they felt that their performance had got completely lost.

From my point of view, the scene illustrated a crucial aspect of the festival as a whole: the artistic quality of any particular dramatic performance was not necessarily the most significant generator of the special atmosphere and experience of the meta-theatrical event. *The Centre of the Campus*, for instance, was a political-aesthetic action that transgressed the usual, indoor, well-defined boundaries of the festival, insinuating that the co-communities might even move on from the plaza into the street; and what was that day only an 'as-if' representation could become the next day a resistant 'as-is' presentation. The performance, with all its artistic weaknesses, managed nevertheless to animate the entrance to the building and attracted people from the neighbourhood who otherwise would not have come, creating a unique and unexpected welcome for those who were hurrying into the building in time to catch the full performance in the big hall. But there another surprise awaited them. None of the groups appeared to be bothered by the official time table, especially those that were supposed to present only 10–15 minutes of their work. Instead they employed all kinds of guerilla tactics to occupy the stage for a little longer.

Baldness in Live Time by the Theatre Group of the Oncology Ward at Hadassah Hospital in Jerusalem, was presented through two short scenes, as formally planned and noted in the programme. The first scene opened with a cheerful dance of little girls with plaits and pink baby-doll outfits. They then disappeared and, in total contrast, little girls in grey robes, whitened faces and with shiny plastic-covered bare skulls' heads, slowly danced around another little girl in a pink baby-doll outfit with face and head similar to theirs. The grey girls then connected themselves to the pink girl by colourful strings. This visual metaphor for the process of the illness was continued in the second act, when the father of the little pink girl sings to her, personifying her baldness:

> When you will learn to love me you will love yourself also. I may be a baldness in a remote wood, but I am your baldness. [...] Don't forget that I came first when you were born. I am the baldness who loves you more than anybody, so smile, smile, smile.

The little pink girl responded by smiling and dancing with the other pink girls. This performance associatively evoked *To Smile*, the previous performance, creating a continuum that reinforced the therapeutic interpretation of the event and the belief in the healing powers of creating theatre and performing it in public. At the end of *Baldness in Live Time*, when the next group was about to hurry on stage in order to rearrange it, Erez Meshulam, the director of *Baldness* 'captured' the stage, asking for a few more minutes. The group, as he said, comprised young Jewish and Arabic patients and their families, and as the Arabs from East Jerusalem did not speak Hebrew but had come to the festival with all their friends and relatives, we, the audience should let them too go on and perform their scene. These three words – Arab, sick and children – constituted an irresistible weapon, and so the Arabic group took the stage for another 15 minutes. This delayed the next performance and stole some of its glory as a full performance, but it was striking nevertheless. An Arab female social worker explained in Hebrew with a heavy Arabic accent, that the song we were going to hear had been written and composed by a Syrian artist, who had had 'the illness' himself and had been cured. A young boy, surrounded by his large family, then opened his mouth and the auditorium was filled with the sound of an amazing voice, singing emotionally in Arabic, leading the entire audience to begin clapping fervently. That was an especially emotional moment, a rare metaphor of the complex Jewish-Arab relationship provided from its somewhat more hopeful side, which only a communitas of co-communities could bring into being.

After such a climax, *Bitter Chocolate*, a performance by the Golden Age Community Theatre from Jaffa C', was received less enthusiastically despite its artistic quality, which was fully deserving of its being performed in entirety. The play, based on real life materials of the elderly actresses, was edited by Michal Revach, a community-based theatre playwright, and directed by Hannah Vazana-Greenvald,[9] a community-based theatre director who has recently been working in particular with groups of women. In the middle of the stage stood a brown bench with the stump of a tree trunk leaning over it. The big, brown abstract painting that had hung on the backdrop during the previous performance was still there. Had it been forgotten? Whose stage design was it originally? Somehow it had become suitable for both the performances, as it alluded perhaps to extracted roots, or to some mysterious, scary illness or creature. On this bench, three old ladies in white gowns were waiting for the Angel of Death, who for some reason was not coming, or perhaps was already there (indicated by the painting?) In the meantime, like Beckett's characters, they were annoying one another and playing with their memories. Each of the women, the happy woman, the snob and the irritable woman, represented a different approach to old age, suggesting the happy, optimistic attitude as the beneficial model. While waiting, they exposed us to some of the crucial moments in their lives, depicted in episodes presented from either side of the stage. In these episodes the three old women – with the audience – watched other women from the group acting out their own characters. In addition to the competent directing and the credible acting, this double witnessing by the old women of other actresses portraying parts of their own lives carried a special theatricality.[10] This device not only heightened self-reflexivity among the audience, but also acted as a pointer that illuminated these women's message for us about the taste of life, which may perhaps not be that of honey, but nonetheless still has the taste of chocolate...

After *Bitter Chocolate* the festival reached its peak in two collages of performance sections that, because of poor timing, were performed almost simultaneously in the big hall on the upper floor and in the smaller hall on the middle floor. Yossi Alfi opened the first sequence with a greeting to the audience, thanking the mayor and all those people who had voluntarily helped to organize the festival, and with a complaint about the difficulties of putting it together. He presented the collage as a 'light tasting' that should encourage us to go on and watch the full performances in their original locations. Alfi briefly introduced each performance and gave the names of the actors. The four different short demonstrations, according to his hegemonic stand, revealed 'the beauty of this country because of the differences'. This intrigued me and I set out to trace the implied co-communities' accumulating politics of difference through the creation of the collage. *Surviving*, by the Women's Community Theatre from Bat Yam, directed by Meyira Medina, depicted the daily life of certain actresses, mainly single mothers who had immigrated from Ethiopia and the former USSR. The play presented the meetings of these women with the social worker, who persistently instructed them on how to become quickly assimilated. As these encounters were presented from the standpoint of the women, we were exposed to those moments of distress that the national, glorious *aliya* (immigration to Israel) narrative usually excludes.

Lena cannot separate herself from her piano, which she was not allowed to take with her. Her piano, personified by another Russian woman, reminds her of her symbiotic past with it, and also symbolizes the mental and cultural anguish of leaving behind such a large part of selfhood, identity and memory. Ziva and Ania demonstrate how difficult it is for two totally different ethnic groups to live in a crowded transit camp, a fact that the Israeli establishment still fails to recognize. Eva, apparently, had to pay a high price to immigrate. Leaving behind her gentile husband, she was now fighting to bring over her only son, who was torn between his parents. Her neighbour, a veteran Mizrahi woman, supported her with unceasing and unquestioning devotion, modelling the kind of reception and relationship we should all display toward newcomers. Medina, the director, who adopted a sparse, stylized aesthetics, seated the actresses, all in black, in an open square of chairs, at front stage. Left stage was a pile of brown suitcases, which besides symbolizing newcomers also functionally transformed into a wall, a table or a counter as required. The action was positioned in the middle of the square, mainly based on the direct acting of the women. The real social worker of the group also played the fictional social worker, and although she may have done so in order to facilitate the role, it turned out to be a subversive tactic exposing the insensitivity of the establishment's 'good will' in their approach to newcomers. The social worker, a former Russian immigrant herself, played her role with total identification. In her rough, Russian-accented Hebrew she preached:

> You see, Israelis are sympathetic neighbors; they are warm and friendly people. If they are sometimes impatient, that can be understood as they have received so many newcomers in recent years. It depends only on you, on how quickly you will assimilate in Israeli society. My suggestion is that you speak only Hebrew, stop babbling Russian or Amharic, speak only Hebrew (becoming ecstatic). Yes, Hebrew only Hebrew, Hebrew! Don't say it is a difficult language; say it is a beautiful language!!!

Having totally internalized her institutional job, she sounded on stage like the blaring trumpet of the hegemonic voice. Was she aware of the ironic, critical viewpoint of the play? Had she joined forces with her group in order to expose the fallacies of the welfare policy? Or had she become an active part of the group in order to control and censor it?

The second performance, *The Past is Dead*, was played by an Ethiopian youth community-based theatre group who boldly presented the intergenerational problem from the adolescents' point of view. For them the past is dead, and they feel totally alienated from their parents' traditional way of life. Another youth theatre, from a more affluent town, presented a scene from *Screens*, exposing the troubles of youngsters who outwardly seem to have the perfect life. The sequence ended in a humorous scene from *In Gerta's Salon* by the Community Theatre of Jaffa D', which parodically portrayed an encounter between a bunch of Ashkenazi hotshots and a shrewd Mizrahi fortune-teller. The formation of the different performance sections into a collage resulted in a kaleidoscopic, critical image of Israeli society from 'the bottom', an image that would be expanded in the subsequent collages. The uniqueness of the second collage lay in its juxtaposition of a group of battered women, a group of battering men and a mixed group. The composition of these three different elements stimulated a more complex sensibility to the problem of domestic violence. Through stylized movements manipulating pieces of iron railing, the women demonstrated the implied violence in every woman's life course, while the men managed to arouse empathy by proving so 'normal' on the one hand, and performing their tormented inner feelings of self-accusation, on the other. The 'eventness' of the collages was engendered by the various spontaneous, celebratory unrehearsed endings to each performance in which family members and friends offered flowers to the actors, or the actors thanked the director and called him/her up on the stage to embrace. The most striking phenomenon however, was the circular physical movement of the theatre groups from the auditorium onto the stage and then back down again to the auditorium. This created a literal/symbolic transformation from spectator to actor and back to spectator, which reached its peak when, at the end of each collage, Alfi invited all the actors from the various performances to come on stage. This process 'elevated' the whole event, generating a flow of 'heightened celebratory consciousness' that reinforced the euphoria that usually accompanies communitas (Rinzler and Seitel, 1982).

Moreover, this form of democratic playing culture signalled the empowering abilities of theatre as an event in general and the festival as a meta-theatrical event in particular.

The second day of the festival repeated the deep structure of the previous day, starting with a low intensity event down in the basement and then building it up with collages in the upper halls. In the storytelling workshop conducted by Yossi Alfi and Igal Azarati, Yossi presented his vision that while

> in the past storytelling belonged to the children, today it has passed to the adults. Theatre has changed to a story-theatre. Everybody tells stories and along the way we see this happening. Everybody loves to tell stories, so we have to use it as a chief method in community-based theatre, help the community to become a storyteller. We don't need artistic skills so much; something that always bothers us concerning community-based theatre. What we need as community directors is to cause the whole community to speak, and find the crucial common denominator. These are the roots of communal drama.

He then guided us to spontaneously tell stories and skilfully connected the stories of the youngsters, who were telling about their journeys all over the world, and the adults, who were occupied with their own or their relatives' immigration stories. The implied common theme was no less than the slippery nature of identity, which, by the end of the day, I had realized was what the festival was all about: various co-communities confronting their socially negated or imposed identity, searching, in turn, for their imagined, more incorporated identity.

The core events of the second day were two collages, one presented by two youth companies and the second by two mentally disabled groups. The first collage was associatively connected to the youth performances of the previous day, and together they focused on adolescents as a temporary co-community that strives to ascertain its self-identity in confrontation with the block-power of adults. The second collage enabled us to perceive something about life and ourselves from the viewpoint of the mentally disabled, and was one of the festival's major socio-aesthetic achievements. In *Similar*, a theatre group from a hostel for the disabled, guided by Osnat Raephaeli-Satobi, a community-based theatre practitioner, presented us with a picture of their independent and communal daily life. The aesthetic scheme, through which the actors personified different machines and activities through pantomime and stylized movements, aroused great applause. The play focused on a young woman with special needs, who unwillingly quits her home because of her mother's severe illness and enters sheltered housing. The extraordinary patience, tolerance, love and support she gets from the residents eventually changes her resistance to accepting herself, and she happily joins the commune of others like her:

> You'd better stay with us. We are all friends here; I'll introduce you to all of them. We have a swimming pool, disco, classes of cooking, painting, Yoga... we'll take you with us. We'll teach you to use the washing machine and the kitchen. You will never be sad again; you will be with us, your friends. It will be fun, you'll see. We'll help you to get used to it. See, we have prepared a party for you, come dance with us (Efrat is surprised but eventually joins the dance).

In *To Learn to Live with it*, the disabled from a special culture centre, guided and directed by Chaim Tal, further develop their critical gaze at us, and offer a more flexible and less overprotective approach to themselves. It was indeed inspiring to see the creative powers of the disabled, who so persuasively managed to impersonate 'normal' and mentally disabled characters. Mr and Mrs Osher (Happiness), a normal couple, give birth to twin sons, one healthy and the other mentally disabled. A distinguished doctor lectures them on their son's problem in a very sophisticated medical language that they barely understand. We witness the pattern of exclusion that this son suffers in each stage of his life – for his own sake and safety of course, as his mother assures him. This continues until the moment when he finds the personal courage to bring an end to the over-protection and goes to live in a hostel with his girl friend.

> I'm sick of being at home, I'm so bored! I always fight with you. Damn it, I'm thirty years old and you still tell me what to do. Don't worry so much, I know how to take care of myself. I'm fed up that so many things my twin brother can do you won't let me do. Enough of it, I'm going to live in a hostel and you will have to learn to live with it.

The successful acting of both 'normal' and 'not-normal' characters by those who have been socially defined as mentally disabled
reduced the gap we usually envisage between these two categories,
indicating that it is maybe we who are acting abnormally in being
so threatened by people who are not exactly like us. It is not only
the mentally disabled but also we who have to 'learn to live with it'.
Tal used a functional staging and the casting was in accordance with
each actor's abilities, which proved to bear aesthetic results as well.
The characters were interesting, humorous and self-reflexive, revealing the power of theatre to confront social stigma and oppressive
mechanisms. The stage design was based on a few big painted cubes
that were transformed into a bed, a hospital bed, a garden bench
and decorated columns in a function hall. The disabled actors, who
had themselves painted these cubes and the placards that were stuck
on them according to each scene, were also the stage managers and
arranged the set.

The two performances attempted to propose a society that transgresses the liberal welfare policy toward the handicapped and instead
generates a more reciprocal interaction between all kinds of people,
recognizing the contribution that the naivety, ease and directness of
stigmatized groups makes to all our lives.

Analysis of the 70 questionnaires about the audience and its reception process reveals that 90 per cent of the spectators were actors,
their relatives and friends, community-based theatre practitioners,
social and community workers and theatre students. The others
were theatregoers who had seen the flyers posted in the neighbourhood streets or heard the advertisement on the local radio. The two
main groups that could be distinguished – 'insiders' and 'outsiders' –
related to the nature of community-based theatre and the experience
of the festival. Both groups described community-based theatre very
similarly as a 'practice that gives voice to invisible communities', that
'is based on people of the community', that 'expresses issues and way
of life that are not represented in the dominant culture' and in which
'the community actors are natural and sometimes even better than
the professionals'. 'Community-based theatre is a liberating theatre',
'a means for confronting social problems', it 'facilitates personal
expression' and 'is a form of protest as well as therapy'. Twenty per
cent of the 'insiders' commented that not all the performances of
the festival were of community-based theatre, and thus on one level

the festival operates to illuminate the ongoing debate about the definition of community-based theatre. Both 'insiders' and 'outsiders' articulated their personal experiences in words such as 'fascinating', 'interesting', 'enjoyable', 'authentic', 'penetrating', 'emotional' and 'exciting and touching'. Forty per cent of the 'insiders' also said that the festival 'empowered them and their choice to work as theatre facilitators', 'it was most impressive to meet other groups that had gone through similar personal, group and creative processes' and 'I am proud of all the groups and their display in the Festival.'

A festival is in fact that institutional performative genre by which the establishment delivers its power and centrality. The festival of theatre in co-communities was a public event that brought together co-communities that, while following the rules of the game, were able to produce an experience of power and centrality that despite its temporary nature was nevertheless transformative; for self and society are generated as they are expressed (Bruner, 1984) and thus every articulation is also a change. Moreover, the festival provided the participants with a means to imagine themselves powerful, united and significant. The festival thus presented a concentrated example that modelled the power of the powerless to appropriate the festival as a cultural institutional practice and to make do with it – for their own benefit – signalling that the staging of an alternative reality can always transgress the fiction and 'restage' reality itself.

Notes

Introduction

1. See Thompson and Schechner (2003, p. xviii).
2. See, for example, Fisher-Lichte (1997, pp. 340–1).
3. The first to draw attention to the 'eventness' of theatre were the painters, sculptors and musicians who, wishing to radicalize their art by theatricalizing it created a new art form, which they called 'the new theatre'. Reclaiming the idea of event, Schechner further developed it into the theatrical event model of his environmental theatre and of the leading groups of the alternative theatre. The theatrical event is an actual, here and now, non-literal, social, sensorial, casual and celebratory occurrence without professional performers. See Shank (2002 [1982]) and Kirby (1965, 1974).
4. The situation I describe here is repeated in other countries as well. See Prentki and Selman (2000) and van Erven (2001a, 2001b).
5. Also called community development and organization.
6. See, for example, Orbe (1998).

1. Dramatic Playing as a Tactic for Confronting the Mask of Ageing

1. See, for example, Myerhoff (1978, 1992).
2. See, for example, Schwartzman (1978), Klinger (1971), Csikszentmihalyi (1975, 1990) and Rapp (1982).
3. See Douglas (1983).
4. See, for example, Eyal (1998).

2. Performing the Scroll of Esther: Articulating Power through Symbolic Inversion

1. See, for example, Chambers (1963 [1903]), Burke (1978) and Bakhtin (1968).
2. See, for example, Burgess (1950), Cummings and Henry (1961) and Anderson (1972).
3. On 'poaching poetics', see the Introduction.
4. See, for example, Schechner (1977).
5. See, for example, Boal (1979, 1992, 1998).
6. See Bateson ([1955] 1976).
7. See, for example, Jung (1933), Bakan (1966) and Gutmann (1987).

3. The Three Elderly Musketeers and their Invention of Play

1. See, for example, Moore and Myerhoff (1977).
2. See, for example, Hall and Neitz (1993).
3. See, for example, Grimes (1982).
4. See, for example, Hazan (1980).
5. On play, ritual and theatre see, for example, Huizinga (1955 [1938]) and Turner (1967, 1982, 1984).

4. Playing the World-Upside-Down at a Children's Medical Centre

1. For example, one of the sword dances was a type of game dance representing a struggle between two groups. The clown, called the Fool, wore a hairy cape with a brush attached to it as a tail, while his part was to perform not the Fool or an animal but the bursar (Chambers, 1963, p. 193).
   Another example is the Revesby Play, another dance-game, which developed into a show with dances. The Fool, dressed as a fool, performed both the presenter and a character in the plot. He fought a 'Hobby-horse' as if it were a horrible monster. After his victory, instead of being rewarded he was punished and put to death by his sons. Pickle Herring, another character in the play, for no logical reason then brought him back to life by stamping his foot, an act which in reality may create some noise, but has nothing to do with either medicine or even witchcraft. This part of the show was replayed several times, followed by a series of sword dances unconnected to the plot. Finally, the Fool and his sons courted a man dressed as a woman named Cecily (208).
2. In recent years hospitals have also welcomed the participation of medical clowns in the daily routine of children's wards.
3. Interpellation is the process by which, according to Louis Althusser, the cultural institutions constitute the individual as a subject who internalizes the dominant ideology and the status quo as natural (Althusser [1969] 1997).

5. 'Theatre of the People': Rhetoric versus an Apparatus for Subversion and Control in the Mizrahi Co-Community

1. See, for example, Bernstein (1989).
2. See, for example, Yishai (1989).
3. All quotes are from the script provided by Yossi Alfi. Community plays in Israel have not yet been published.
4. See, for example, Shohat (2001).
5. All quotes are from the script provided by Igal Azarati.

6. See, for example, Hever et al. (2002).
7. All quotes are from the script provided by Peter Harris.

6. Battered Women on the Stage: from Spoken Objects to Speaking Subjects

1. As reported by the Israeli daily newspaper *Yedioth Ahronot*, 25–26 November 2001.
2. When the play was first performed, Neve Zedek was a disadvantaged neighbourhood of Tel Aviv, inhabited mostly by Jews of Yemenite origin. Over the years it has become a centre for recreation and artistic creation.
3. On the documentary theatre of Nola Chilton see Ben-Zvi (2007).
4. On the Mizrahi ethnic problem, see the previous chapter.
5. The quotes are from the unpublished printed script in the theatre archive of Tel Aviv University.
6. All quotes are from the script provided by Hannah Vazana-Greenvald.
7. Mulvey (1975), Goodman (1996) and Case (1988).

7. Between Home and Homeland: Ethiopian Youth Making Do with Theatre in a Boarding School

1. See, eniar.org (accessed December 2009).
2. Chen Elia was at that time a student in my course 'Gender, Class and Ethnicity in Israeli Community-Based Theatre', at the Theatre Department in Tel Aviv University.
3. All the information given here is taken from the notebook of Chen Elia and I thank her for her cooperation.
4. These excerpts are from the notebook of Chen Elia.
5. On this subject, see for example, Lev-Aladgem (2003b) and Lev-Aladgem and First (2004). For a discussion of this performance see Chapter 5.
6. 'Spect-actors' is Boal's term for participants who are both spectators and actors.
7. All quotes are from the text provided by Chen Elia.

8. Undoing Political Conflict: Israeli Jews and Palestinians Co-Creating a Theatrical Event

1. Also Ramle and Ramlhe.
2. Yussuf Swaid is now a noted theatre and television actor in Israel.
3. Family 'honour-killing' is a traditional phenomenon in Islamic communities in countries such as Bangladesh, Brazil, Egypt, India, Jordan, Israel, Great Britain, Italy, Ecuador, Pakistan, Morocco, Sweden, Turkey and Uganda. This violent 'duty' is expected of a male relative, who has to kill the woman who has behaved improperly and shamed her family.

4. The Law of Return (1950) gives every Jew the right to immigrate to Israel and automatically acquire citizenship. This right is completely denied the Arab refugees who left the country during the 1948 war.
5. On 'emic analysis,' and 'thick description' see, for example, Geertz (1973).
6. On the subject of mother as oppressor see for example, Chapter 6 above and Lev-Aladgem and First (2004).
7. All quotes are from Tali and Mizrahi-Shapira (2005).

9. The National Festival of the Co-Communities: the Meta-Theatrical Event

1. During this year Alfi ceased to administer the Givatayim Theatre but managed to organize the festival for one more year, after which he was no longer able to raise the large budget that the theatre required.
2. See, for example, Chapter 5 above and Lev-Aladgem (2003b).
3. During the 1980s he organized workshops for community-based theatre directors in various academic venues through 'Kahal' (in English, 'audience'), his own non-profit organization.
4. Levana is the official secretary of 'Kahal', faithfully assisting Alfi for many years.
5. I myself was a member of the organization board of the first festival in 1998 and led discussions at the festivals of 1999 and 2003. As I had chosen to focus as a researcher on the 2002 festival, I refrained from taking any active part.
6. A few community performances are also selected each year to participate in Theatrical Autumn, a festival for amateur theatres organized by the Tel Aviv municipality. At this festival each performance is fully presented.
7. See, for example, Turner (1967) and Turner and Turner (1982).
8. On Igal Azarati, see also Chapter 5.
9. On Hannah Vazana-Greenvald, see also Chapter 6.
10. On this subject see, for example, Rokem (2002).

Bibliography

Adams, David Wallace, *Education for Extinction: American Indians and the Boarding School Experience 1875–1928*, Lawrence: University of Kansas Press, 1995.

Alfi, Yossi, *Theater and Community*, Jerusalem: Jewish Agency for Israel Renewal Department Planning Unit, 1986 (Hebrew).

—— Interview with Shulamith Lev-Aladgem, Tel Aviv, 12 August 2002 (unpublished).

Althusser, Louis, 'Ideology and State Apparatuses', in Antony Easthope and Kate McGowan (eds), *A Critical and Cultural Theory Reader*, Toronto: University of Toronto Press, [1969] 1997, pp. 50–8.

Anderson, Barbara, 'The Process of Deculturation: its Dynamics among United States Aged', *Anthropological Quarterly*, 45, 1972: 209–16.

Anderson, Benedict, *Imagined Communities*, 2nd revised edition, London: Verso, 1991.

Appadurai, Arjun, *Modernity at Large: Cultural Dimensions of Globalization*, Minneapolis and London: University of Minnesota Press, 1996.

Archer, Margaret M., *Being Human: the Problem of Agency*, Cambridge: Cambridge University Press, 2000.

Ardener, Shirley, *Perceiving Women*, London: Malaby, 1975.

—— *Defining Females: the Nature of Women in Society*, New York: John Wiley, 1978.

Atlan, N. Henri, *A Tort et a Raison: Intercritique de la Science et du Mythe* (Hebrew translation by Dan Daor), Tel Aviv: Am Oved, 1994 [1968].

Azarati, Igal, Interview with Shulamith Lev-Aladgem, Tel-Aviv, 12 August 2001 (Hebrew) (unpubished).

—— Interview with Shulamith Lev-Aladgem, Tel Aviv, 20 May 2003 (Hebrew) (unpublished).

Babcock, Barbara (ed), *The Reversible World: Symbolic Inversion in Art and Society*, Ithaca and London: Cornell University Press, 1978.

Bakan, David, *The Duality of Human Existence: Isolation and Communication in Western Man*, Boston: Beacon Press, 1966.

Bakhtin, Mikhail, *Rabelais and His World* (trans. Helene Iswolsky), Cambridge, MA: MIT Press, 1968.

—— *Problems of Dostoyevsky's Poetics* (trans. R.W. Rotsel), Ann Arbor, MI: Ardis, 1973.

Bar-Tal, Daniel, 'Socialization into Conflict: a General Perspective', in Yechezkel Rachamim and Daniel Bar-Tal (eds), *Socialization into Conflict in Israeli-Jewish Society*, Tel Aviv: Walter Lebach Institute of Jewish-Arab Coexistence through Education, 2006, pp. 15–22.

Bateson, Gregory, 'A Theory of Play and Fantasy', in Jerome Bruner, Alison Jolly and Kathy Sylva (eds), *Play – Its Role in Development and Evolution*, New York: Basic Books, 1976 [1955], pp. 119–29

Baudrillard, Jean, *Symbolic Exchange and Death*, London: Sage, 1993 [1976].

Beck, Ulrich, '"Zombie Terms". An Interview with Ulrich Beck', *Theory and Criticism*, 16, 2000: 247–62 (Hebrew).

Ben-David, Amit and Adital Tirosh Ben Ari, 'The Experience of Being Different: Black Jews in Israel', *Journal of Black Studies*, 27(4), 1997: 510–18.

Ben-Zvi, Linda, 'Staging Calamities of Separation: the Documentary Theatre of Nola Chilton', in Linda Ben-Zvi and Tracy Davis (eds), *Considering Calamity: Methods for Performance Research*, Tel Aviv: Assaph, 2007, pp. 129–47.

Bernstein, Devora, 'The Black Panthers: Conflict and Protest in Israeli Society', in Moshe Lisak (ed), *Stratification in Israeli Society: Ethnic, National and Class Cleavages*, Tel Aviv: The Open University, 1989, pp. 591–606 (Hebrew).

Bevers, Tom, Peter van den Hurk and Femke van Schie (eds), *International Community Theatre Festival: a Comprehensive Report*, Rotterdam: Rotterdams Wijktheater, 2001.

Bhabha, Homi, *The Location of Culture*, London: Routledge, 1994.

Boal, Augusto, *Theatre of the Oppressed* (trans. Charles A. and Maria-Odilia Leal McBride), London: Pluto Press, 1979.

―――― *Games for Actors and Non-Actors* (trans. Adrian Jackson), London and New York: Routledge, 1992.

―――― *Legislative Theatre* (trans. Adrian Jackson), London and New York: Routledge, 1998.

Brady, Sara, 'Welded to the Ladle: *Steelbound* and Non-Radicality in Community Theater', *Drama Review*, 44(3) (T167) 2000: 51–74.

Braziel, Jana Evans and Anita Nannur (eds), *Theorizing Diaspora*, Malden, MA: Blackwell Publishing, 2003.

Brook, Peter, *The Empty Space*, Harmondsworth: Penguin, 1968.

Bruner, Edward (ed), *Text, Play, and Story: the Construction and Reconstruction of Self and Society*, Long Grove, IL: Waveland Press, 1984.

Burgess, W. Ernest, 'Personal and Social Adjustment in Old Age', in Milton Derber (ed.), *The Aged and Society*, Champaign: Industrial Relations Research Association, 1950, pp. 138–256.

Burke, Peter, *Popular Culture in Early Modern Europe*, London: Temple Smith, 1978.

Butler, Judith, *Gender Trouble: Feminism and the Subversion of Identity*, London: Routledge, 1990.

Caillois, Roger, *Man, Play and Games*, New York: Free Press, 1961.

Case, Sue-Ellen, *Feminism and Theatre*, London: Routledge, 1988.

Chambers, K. Edmund, *The Medieval Stage*, London: Oxford University Press, 1963 [1903].

Chinman, Matthew. J. and Jean A. Linney, 'Toward a Model of Adolescent Empowerment: Theoretical and Empirical Evidence', *Journal of Primary Prevention*, 18(4), 1998: 393–413.

Coser, Lewis A., *Greedy Institutions: Patterns of Undivided Commitment*, New York and London: Collier Macmillan Publishers, 1974.

Csikszentmihalyi, Mihaly, *Beyond Boredom and Anxiety*, San Francisco, Washington and London: Jossey-Bass, 1975.

——— *The Psychology of Optimal Experience*, New York: Harper and Row, 1990.

Cummings, Elaine and William Henry, *Growing Old: the Process of Disengagement*, New York: Basic Books, 1961.

Dawson, Gary Fisher, *Documentary Theatre in the United States*, Westport, CT: Greenwood, 1999.

de Certeau, Michel, *The Practice of Everyday Life*, Los Angeles and London: University of California Press, 1984.

Deleuze, Gilles, 'Nomade Thought', in David B. Allison (ed.), *The New Nietzsche*, New York: Dell Publishing, 1977, pp. 142–9.

Deleuze, Gilles and Felix Guattrai, 'Qu'est-ce Qu'une Literature Mineure?' (trans. Orly Azuli), *Mican*, 1, 2000: 134–43 (Hebrew).

Delvin, Diana, *Mask and Scene*, London: Macmillan, 1989.

Diamond, Elin, 'Mimesis, Mimicry and the True-Real', *Modern Drama*, 32, 1989: 58–72.

Dolan, Jill, *The Feminist Spectator as Critic*, Ann Arbor, MI: University of Michigan Press, 1988.

Douglas, Mary, *Natural Symbols: Explorations in Cosmology*, New York: Vintage Books, 1973.

Douglas, Tom, *Groups: Understanding People Gathered Together*, London and New York: Tavistock Publications, 1983.

Eco, Umberto, V.V. Ivanov and Monica Rector, *Carnival!*, ed. Thomas A. Sebeok, Approaches to Semiotics 64, Berlin, New York and Amsterdam: Mouton, 1984.

Elia, Chen, Interview with Shulamith Lev-Aladgem, Tel-Aviv, 13 March 2007 (unpublished).

Eyal, Nitza, 'The Personal Experience of Time of the Aged', *Gerontology*, 25(3–4), 1998: 9–24 (Hebrew).

Favorini, Attilio, 'Representation and Reality: the Case of Documentary Theatre', *Theatre Survey*, 35(2), 1994: 31–43.

Featherstone, Mike and Mike Hepworth, 'The Mask of Aging and the Postmodern Life Course', in Mike Featherstone, Mike Hepworth and Bryan S. Turner (eds), *The Body: Social Processes and Cultural Theory*, London: Sage, 1991, pp. 370–89.

Fink, Eugen, 'The Oasis of Happiness: Toward an Ontology of Play', *Yale French Studies*, 41, 1968: 19–30.

First, Anat and Eli Avraham, 'Changes in the Political, Cultural, and Media Environment and their Impact on the Coverage of Conflict: the Case of the Arab Population in Israel', *Conflict and Communication*, 2003, 2(1): 1–14.

——— 'The Good, the Bad and the Absent: Contradictory Trends in the Treatment of Arabs on Israeli TV', *Mediterranean Journal of Human Rights*, 8(2), 2004: 55–78.

Fisher-Lichte, Erika, *The Show and the Gaze of the Theatre*, Iowa: University of Iowa Press, 1997.

Fox, Pamela, *Class Fictions: Shame and Resistance in the British Working-Class Novel 1890–1945*, Durham, NC: Duke University Press, 1994.

Geertz, Clifford, *The Interpretation of Culture*, New York: Basic Books, 1973.

Georgiou, Myria, *Diaspora, Identity and the Media*, Kresskill, NJ: Hampton Press, 2006.

Goffman, Erving, 'Characteristics of Total Institutions', in Maurice R. Stein, Arthur J. Vidich, and David Manning White (eds), *Identity and Anxiety: Survival of the Person in Mass Society*, New York: The Free Press, 1960, pp. 449–79.

—— *Asylums: Essays on the Social Situations of Mental Patients and Other Inmates*, Garden City, NY: Anchor, 1961.

Goodman, Lizbeth, 'Feminisms and Theatres: Canon Fodder and Cultural Change', in Patrick Campbell (ed.), *Analyzing Performance: a Critical Reader*, Manchester and New York: Manchester University Press, 1996, pp. 19–42.

Gramsci, Antonio, *Selections from the Prison Notebooks* (trans. Alon Altaras), Tel Aviv: Resling, 2004 (Hebrew).

Graves, Benjamin, 'Homi K. Bhabha: the Liminal Negotiation of Cultural Difference', Political Discourse – Theories of Colonialism and Postcolonialism, Brown University, http://www.postcolonialweb.org/poldiscourse/bhabha/bhabha1.html, accessed December 2009.

Grimes, Ronald, *Beginnings in Ritual Studies*, New York: University Press of America, 1982.

Grossman, David, *Present-Absentees*, Tel Aviv: Hakibbutz Hameuchad, 1992 (Hebrew).

Grotowski, Jerzy, *Toward a Poor Theatre*, London: Methuen, 1969.

Gutmann, David, *Reclaimed Powers: Toward a New Psychology of Men and Women in Later Life*, New York: Basic Books, 1987.

Hall, R. John and Mary Jo Neitz, *Culture: Sociological Perspectives*, Englewood Cliffs, NJ: Prentice Hall, 1993.

Hall, Stuart, 'Cultural Identity and Diaspora', in Jana Evans Braziel and Anita Mannur (eds), *Theorizing Diaspora*, Malden, MA: Blackwell, 2003, pp. 233–46.

Handelman, Don, *Work and Play among the Aged*, Amsterdam: Van Gorcum, 1977.

—— *Models and Mirrors: Toward an Anthropology of Public Events*, Cambridge: Cambridge University Press, 1990.

Harris, Peter, Interview with Shulamith Lev-Aladgem, Tel-Aviv, 5 June 2003 (Hebrew) (unpublished).

Hazan, Haim, *The Limbo People*, London: Routledge, 1980.

—— *Aging as a Social Phenomenon*, Tel Aviv: Ministry of Defense, 1988 (Hebrew).

—— *Old Age: Constructions and Deconstructions*, Cambridge: Cambridge University Press, 1994.

—— 'Age', in Uri Ram and Nitza Berkowitz (eds), *Inequality*, Beer Sheva: Ben Gurion University Publications, 2006a, pp. 82–9 (Hebrew).

—— 'Beyond Discourse; Recognizing Bare Life Among the Very Old', in Jason L. Powell and Azrrini Wahidin (eds), *Foucault and Aging*, New York: Nova Science Publishers, 2006b, pp. 157–70.

———— 'Essential Others: Anthropology and the Return of the Old Savage', *International Journal of Sociology and Social Policy*, 29(1/2), 2009: 60–72.

Hever, Hannan, Yehouda Shenhav and Pnina Motzafi-Haller (eds), *Mizrahim in Israel: a Critical Observation into Israel's Ethnicity*, Jerusalem: Van Leer Institution and Hakibbutz Hameuchad, 2002 (Hebrew).

Hoare, Quintin and Geoffrey Nowell-Smith (eds), *Selections from the Prison Notebooks of Antonio Gramsci*, New York: International Publishers, 1971.

Hobsbawm, Erik and Terrence Ranger (eds), *The Inversion of Tradition*, Cambridge: Cambridge University Press, 1983.

Huizinga, John, *Homo Ludens: a Study of the Play-Element in Culture*, Boston: Beacon, 1955.

Huppart, Miriam, 'Community Theater in an Urban City', in Yossi Alfi (ed.), *What is Community Theater?* Ramat-Gan: Bar Ilan University, 1975, pp. 28–30 (Hebrew).

Huxley, Julian, 'The Courtship Habits of the Great Crested Grebe (*Podicepts Cristatus*), with an Addition to the Theory of Sexual Selection', *Proceedings of the Zoological Society*, 35, 1914: 491–562.

Jamal, Amal, 'Nationalizing State and the Constitution of "Hollow Citizenship": Israel and its Palestinian Citizens', *Ethnopolitics*, 6(4), 2007: 471–93.

Jenkyns, Marina, *The Play's the Thing*, London: Routledge, 1996.

Joseph, Haya, 'Between the Hammer and the Anvil', *Chotam*, July 1981 (Hebrew).

Jung, C.G., *Modern Man in Search of a Soul*, New York: Harcourt, Brace and World, 1933.

Kapferer, Bruce, 'The Ritual Process and the Problem of Reflexivity in Sinhalese Demon Exorcism', in John MacAloon (ed.), *Rite, Drama, Festival, Spectacle*, Philadelphia: I.S.H.I, 1984, pp. 179–208.

Kashti, Yitzhak and Mordecai Arieli, 'Classification of Boarding Schools', in David Nevo (ed.), *The Educational Process, Discussion and Research*, Tel Aviv: Dvir, 1997, pp. 421–30 (Hebrew).

Kastenbaum, Robert (ed), *Old Age on the New Screen*, New York: Springer Publishing Company, 1981.

Katzanelson, Edna, *Koev Li: Mental Confrontation with Pain among Children*, Tel-Aviv: Amichai, 1993 (Hebrew).

Kimmerling, Baruch, *Immigrants, Settlers, Natives: the Israeli State and Society between Cultural Pluralism and Cultural Wars*, Tel Aviv: Am Oved, 2004 (Hebrew).

Kirby, Michael, *Happenings*, New York: Dutton, 1965.

———— (ed.), *The New Theatre*, New York: New York University Press, 1974.

Klinger, Evelyne, *Structure and Function of Fantasy*, New York: John Wiley, 1971.

Kobowitz, Yaniv, 'In Ethiopia I Discovered Baruch Dego', *Haaretz*, 2 October 2007 (Hebrew).

Kohansky, Mendel, 'For Battered, for Worse', *Jerusalem Post Magazine*, 4 September 1981.

Kramarea, Cheris, *Women and Men Speaking*, Rowley: Newbury House, 1981.

Laclau, Ernesto and Chantal Mouffe, *Hegemonic and Socialist Strategy: Toward a Radical Democratic Politics*, London and New York: Verso, 2001.

Lahr, John, *Acting Out America: Essays on Modern Theatre*, Harmondsworth: Penguin, 1972.

Lappin, Rimona, Interview with Shulamith Lev-Aladgem, Tel Aviv, 8 February 2006 (Hebrew) (unpublished).

Latour, Bruno, *We Have Never Been Modern*, Cambridge, MA: Harvard University Press, 1993.

Lavie, Smadar and Ted Swedenburg (eds), *Displacement, Diaspora, and Geographies of Identity*, Durham and London: Duke University Press, 1996.

Lev-Aladgem, Shulamith, 'The Dramatic Play Workshop: the Playful Performance in the Community', dissertation, Tel Aviv University, 1995.

—— 'Drama in a Geriatric Day-Care Center', *Gerontology*, 76, 1996/7: 48–56 (Hebrew).

—— 'Improvisation upon the Scroll of Esther: Symbolic Inversion in an Adult Day-Care Centre', *Journal of Folklore Research*, 35(2), 1998: 127–45.

—— 'From Ritual to Drama and Back in a Day-Care Centre', *Journal of Aging Studies*, 13(3), 1999: 315–33.

—— 'Dramatic Play amongst the Aged', *British Dramatherapy Journal*, 21(3), 2000:3–10.

—— 'Carnivalesque Enactment at the Children's Medical Centre of Rabin Hospital', *Research in Drama Education*, 5(2), 2000: 163–74.

—— 'From Object to Subject: Israeli Theatres of the Battered Women', *New Theatre Quarterly*, XIX(2), 2003a: 139–49.

—— 'Ethnicity, Class and Gender in Israeli Community Theatre', *Theatre Research International*, 28(2), 2003b: 181–92.

—— 'Whose Play is it? The Issue of Authorship/Ownership in Israeli Community Theatre', *Drama Review*, 48(3) (T. 183), 2004: 117–34.

—— 'The Israeli National Community Theatre Festival: the Real and the Imagined', *Theatre Research International*, 30(3), 2005: 284–95.

—— 'Between Home and Homeland: Facilitating Theatre with Ethiopian Youth', *Research in Drama Education*, 13(3), 2008: 275–93.

Lev-Aladgem, Shulamith and Anat First, 'Community Theatre as a Site for Performing Gender and Identity', *Feminist Media Studies*, 4(1), 2004: 37–50.

Lindenberger, Herbert, *Historical Drama: the Revolution of Literature and Reality*, Chicago: University of Chicago Press, 1975.

Lorenz, Konrad, 'Habit, Ritual and Magic' [1916], in Richard Schechner and Mady Schuman (eds), *Ritual, Play and Performance*, New York: Seabury Press, 1976, pp. 18–34.

Mack, Danielle, Interview with Shulamith Lev-Aladgem, Herzelia, 15 August 2001 (unpublished).

Mann, Emily, Interview by Gary Fisher Dawson, New York City, 18 January 1994, in Gary Fisher Dawson, *Documentary Theatre in the United States*, Westport, CT: Greenwood, 1999, p. 23, n. 20.

Marcuse, Herbert, *Five Lectures*, Boston: Beacon, 1970.

—— *The End of Utopia*, Tel Aviv: Am Oved, 1972 (Hebrew).

Masica, Yamin, Interview with Shulamith Lev-Aaldgem, Ramat-Gan, 3 September 2001 (Hebrew) (unpublished).

McCaslin, Nellie, *Creative Drama in the Classroom*, New York: Longman, 1990.

Medina, Meyira, Interview with Shulamith Lev-Aladgem, Tel Aviv, 20 April 2003 (Hebrew) (unpublished).

Michaels, Claire, 'Geriadrama', in Gertrud Schattner and Richard Courtney (eds), *Drama in Therapy*, New York: Drama Books Specialists, 1981, Vol. 2, pp. 175–97.

Miller, Luis, 'Theater and Community', *Bama*, 56, 1973: 82–8 (Hebrew).

—— 'Theatre and Community: Joseph's Tent, *Bama*, 63/64, 1975: 69–77 (Hebrew).

—— 'Theater and Creativity', *Bama*, 73/74, 1977: 85–9 (Hebrew).

Mizrahi-Shapira, Oshrat, Interview with Shulamith Lev-Aladgem, Tel Aviv, 13 February 2006 (Hebrew).

Moore, F. Sally and Barbara G. Myerhoff (eds), *Secular Ritual*, Amsterdam: Gorcum, 1977.

Mulvey, Laura, 'Visual Pleasure and Narrative Cinema', *Screen*, 16(3), 1975: 6–18.

Myerhoff, Barbara, *Number Our Days*, New York: Dutton, 1978.

—— 'A Death in Due Time: Construction of Self and Culture in Ritual Drama', in John. MacAloon (ed), *Rite, Drama, Festival, Spectacle*, Philadelphia: I.S.H.I, 1984, pp. 149–77.

—— *Remembered Lives: the Work of Ritual, Storytelling, and Growing Older*, Ann Arbor, MI: University of Michigan Press, 1992.

Naaman, Adhit, 'To Act Battered Women', *Yedioth Ahronot*, 10 July 1981 (Hebrew).

Naremore, James and Partick Brantlinger, 'Introduction: Six Artistic Cultures', in James Naremore and Patrick Brantlinger (eds), *Modernity and Mass Culture*, Bloomington and Indianapolis: Indiana University Press, 1991, pp. 1–23.

Nicholson, Helen, *Applied Drama*, Basingstoke and New York: Palgrave, 2005.

Ojanuga, Durrenda, 'The Ethiopian Jewish Experience as Blacks in Israel', *Journal of Black Studies*, 24(2), 1993: 147–58

Ophek, David, 'Phachme Dast' (documentary film), Diva Productions, 1997.

Orbe, Mark, *Constructing Co-Cultural Theory*, London: Sage, 1998.

Peled, Yoav and Gershon Shafir, *Being Israeli: the Dynamics of Multiple Citizenship*, Tel Aviv: Tel Aviv University Press, 2005 (Hebrew).

Pelled, Asapha, 'The Black Book', *Yediot Aharonoth*, 7 December 2007 (Hebrew).

Pharan, Michael, Interview with Shulamith Lev-Aladgem, Ramat-Gan, 23 November 2003 (Hebrew) (unpublished).

Prentki, Tim and Jan Selman, *Popular Theatre in Political Culture: Britain and Canada in Focus*, Bristol and Portland: Intellect, 2000.

Rapp, Uri, 'Simulation and Imagination: Mimesis as Play', *Maske und Kothutn*, 28, 1982: 67–86.

Rinzler, Ralph and Peter Seitel, 'Preface', in Victor Turner (ed.), *Celebration: Studies in Festivity and Ritual*, Washington DC: Smithsonian Institution Press, 1982.

Rokem, Freddie, 'Witnessing Woyzeck: Theatricality and the Empowerment of the Spectator', *Substance*, 31(2&3), 2002: 167–83.

Roose-Evans, James, *Experimental Theatre from Stanislavski to Peter Brook*, London: Routledge and Kegan Paul, 1984.

Schechner, Richard, 'Approaches to Theory/Criticism', *Tulane Drama Review*, 10(4), 1966: 20–53.

——— *Public Domain*, New York: Bobbs-Merrill, 1969.

——— 'Actuals: Primitive Ritual and Performance Theory', *Theatre Quarterly*, 1(2),1971: 49–65.

——— 'Toward a Poetics of Performance', in *Essays on Performance Theory, 1970–1976*, New York: Drama Books Specialists, 1977, pp. 109–39.

——— *The End of Humanism*, New York: Performing Art Journal Publications, 1982.

——— 'Playing', *Play and Culture*, 1(1), 1998: 3–19.

——— 'Believed-in Theatre', *Performance Research*, 2(2), 1997: 77–91.

——— *Performance Studies*, London and New York: Routledge, 2002.

Schwartzman, Helen, *Transformations: the Anthropology of Child's Play*, New York and London: Plenum Press, 1978.

Schinina, Guglielmo, 'Social Theatre and Some Open Questions about its Developments', *Theatre Review*, 58(3), 2004: 17–31.

Shabtay, Malka, 'Living with Threatened Identities: the Experiences of Ethiopian Youth in Israel Living with a Color Difference in an Ethnocentric Climate', *Megamot*, 41(1–2), 2001: 97–112 (Hebrew).

Shank, Theodore, *Beyond the Boundaries, American Alternative Theatre*, Ann Arbor, MI: The University of Michigan Press, 2002 [1982].

Shapran, Ziva, Television Channel, *Encounter with Documentary Theatre*, 1976; programme notes to *Bicycle for a Year*, Haifa Municipal Theatre, 1978 (Hebrew).

Shem-Tov, Naphtali, Interview with Shulamith Lev-Aladgem, Ramat-Gan, 28 March 2002 (Hebrew) (unpublished).

Shenhav, Yehuda, 'Jews from Arab Countries: an Ethnic Community in the Realms of National Memory', in Hannan Hever, Yehuda Shenhav and Pnina Motzafi-Haller (eds), *Mizrahim in Israel: a Critical Observation into Israel's Ethnicity*, Tel Aviv: The Van Leer Jerusalem Institution and Hakibbutz Hameuchad, 2002, pp. 105–51 (Hebrew).

——— *The Arab-Jews: Nationalism, Religion and Ethnicity*, Tel Aviv: Am Oved, 2003 (Hebrew).

——— 'Introduction', in Zigmunt Bauman, *Liquid Modernity*, Jerusalem: Magness, 2007 (Hebrew).

Shohat, Ella, *Forbidden Reminiscences*, Tel Aviv: Bimat Kedem, 2001 (Hebrew).

Smooha, Samy, *Israel: Pluralism and Conflict*, Berkeley: University of California Press, 1978.

Smooha, Samy, *Arabs and Jews in Israel, Vol 1: Conflicting and Shared Attitudes in a Divided Society*, Boulder, CO: Westview, 1989.

—— 'Minority Status in an Ethnic Democracy: the Status of the Arab Minority in Israel', *Ethnic and Racial Studies*, 13, 3, 1990: 389–413.

—— 'Class, Ethnic, and National Cleavages and Democracy in Israel', in Ehud Shprinzak and Larry Diamond (eds), *Israeli Democracy under Stress*, Boulder, CO: Lynne Rienner, 1993, pp. 309–42.

Stanislavski, Constantin, *An Actor Prepares*, New York: Routledge, 1989.

Svirski, Barbara, 'A Shelter for Battered Women, an Intermediary Report November1977–March 1978', Centre for Advanced Services, Jerusalem, 1978 (Hebrew).

Swaid, Yussuf, Interview with Shulamith Lev-Aladgem, Tel Aviv, 15 February 2006 (Hebrew) (unpublished).

Tagrin, Yael and Hannah Vazana, 'The Community Theatre Group: a Group Therapy Modula', in *Models for Group Therapy with Women, Men and Children who live in a Violent Environment*, Hertzlia: Published by the Center for Prevention and Treatment of Domestic Violence, 1999 (Hebrew).

Tali, Sa'ad, Interview with Shulamith Lev-Aladgem, 18 February 2006 (Hebrew) (unpublished).

Tali, Sa'ad and Oshrat Mizrahi-Shapira, 'Bridging through Theatre', in Talia Levine Bar-Yoseph (ed.), *The Bridge Dialogues across Cultures*, New Orleans: Gestalt Institute of New Orleans, 2005, pp. 133–48.

Taylor, Philip, *Applied Theatre*, Portsmouth: Heinemann, 2003.

Thompson, James, 'Digging up Stories', *Theatre Review*, 58(3), 2004: 150–64.

Thompson, James and Richard Schechner, 'Why "Social Theatre"', *Drama Review*, 48(3), 2004: 11–16.

Turner, Victor, *The Forest of Symbols*, New York: Cornell University Press, 1967.

—— *The Ritual Process*, Chicago: Aldine, 1969.

—— *From Ritual to Theatre*, New York: P.A.J, 1982.

—— 'Liminality and the Performative Genres', in John MacAloon (ed.), *Rite, Drama, Festival, Spectacle*, Philadelphia: I.S.H.I, 1984, pp. 19–41.

Turner, Victor and Edith Turner, 'Religious Celebrations', in Victor Turner (ed.), *Celebrations: Studies in Festivity and Ritual*, Washington DC: Smithsonian Institution Press, 1982, pp. 201–19.

Urian, Dan, 'Israeli Theatre and the Ethnic Problem: the Case of Nola Chilton', *Assaph*, 16, 2000: 135–56.

van Erven, Eugene, *Community Theatre: Global Perspectives*, London and New York, Routledge, 2001a.

—— 'Community Theatre is Art!' in Tom Bevers, Peter van den Hurk and Femke van Schie (eds), *International Community Theatre Festival: a Comprehensive Report*, Rotterdam: Rotterdams Wijktheater, 2001b, pp. 71–5.

van Gennep, Arnold, *The Rites of Passage*, trans. M.B. Vizedom and G.L. Caffee, Chicago: University of Chicago Press, 1960 [1908].

Wittgenstein, Ludwig, *Philosophical Investigations*, London: Macmillan, 1958.

Yishai, Yael, 'Israel's Right-Wing Jewish Proletariat', in Moshe Lisak (ed.). *Stratification in Israeli Society: Ethnic, National and Class Cleavages*, Tel Aviv: The Open University, 1989, pp. 626–39 (Hebrew).

Zeltzer, Gita, 'The Community Theater: Expectations and Reality', MA thesis, Tel Aviv University, 1986 (Hebrew).

Zerubavel, Yael, *Recovered Roots: the Making of Israeli National Tradition*, Chicago and London: University of Chicago Press, 1995.

Ziv, Melech, *Dos Purim-shpil: Do Shpilt di Role Haman un Mordekhai*, by S. Weissenberg (trans. Melech Ziv), with an Introduction by the translator, Tel Aviv: Peretz Publishing House, 1995 (Hebrew).

Index